Facilities Management

Facilities Management
The Dynamics of Excellence

Third Edition

Peter Barrett

Professor of Management in Property and Construction, School of the Built Environment, University of Salford

Edward Finch

Freelance Facilities Management Commentator and former Professor of Facilities Management, University of Salford

WILEY Blackwell

This edition first published 2014
© 2014 by John Wiley & Sons, Ltd
1st edition: © 1995 Blackwell Science Ltd
2nd edition: © 2003 Blackwell Publishing Ltd

Registered Office
John Wiley & Sons, Ltd, The Atrium, Southern Gate, Chichester, West Sussex, PO19 8SQ,
United Kingdom.

Editorial Offices
9600 Garsington Road, Oxford, OX4 2DQ, United Kingdom.
The Atrium, Southern Gate, Chichester, West Sussex, PO19 8SQ, United Kingdom.

For details of our global editorial offices, for customer services and for information about how
to apply for permission to reuse the copyright material in this book please see our website at
www.wiley.com/wiley-blackwell.

The right of the author to be identified as the author of this work has been asserted in accordance
with the UK Copyright, Designs and Patents Act 1988.

Library of Congress Cataloging-in-Publication Data

Barrett, Peter (Peter Stephen), 1957–
    Facilities management : the dynamics of excellence / Peter Barrett, Edward Finch. – Third edition.
        pages   cm
    Includes bibliographical references and index.
    ISBN 978-0-470-67397-3 (pbk.)
 1. Real estate management.   2. Facility management.   I. Finch, Edward.   II. Title.
    HD1394.B37 2013
    658.2–dc23
                                                            2013019873

A catalogue record for this book is available from the British Library.

Wiley also publishes its books in a variety of electronic formats. Some content that appears in print
may not be available in electronic books.

Cover design by Sandra Heath
Cover image courtesy of Shutterstock and the authors

Set in 10/12.5pt Minion by SPi Publisher Services, Pondicherry, India

1   2014

Contents

Preface

This book is about unlocking the full potential of facilities through excellent facilities management. The innate potential of built space can affect the way that we work, play, heal, learn and generally cope with the demands of modern life and modern business. The burgeoning profession of facilities management can be instrumental in realising this potential. However, poor management can lead to dysfunctional and short-lived built assets misaligned with changing user demands.

The two preceding editions of this book were instrumental in defining the landscape of facilities management (FM). As such, the book itself has become something of a standard reference for professionals and students alike. However, the landscape of FM has changed significantly in recent years, necessitating a further revision of the book (Third Edition). The concepts and ideas espoused in the earlier editions are just as relevant today as when they were first explored. Indeed, it is somewhat surprising just how robust the principles have been. What has necessitated a new edition is not a replacement of previous ideas, but rather a reconnection of established notions with the modern context, shaped as it is by unprecedented changes in the political, social and technological arenas.

The proliferation of good practice guides in facilities management suggests a world in which clarity exists between the right way of doing things and the wrong way. This singularity in thinking is sometimes at odds with the 'systems thinking' developed throughout this book. Indeed, the systems approach suggests that there are generally many ways of achieving the same goal. For many, this proposition is somewhat discomforting. For others, it is stimulating and invigorating. It provides the challenge to enable excellent facilities managers to shine.

Often we are tempted to think of physical facilities as purely economic commodities that we buy and sell in a property market according to our needs. Fundamental to the success of facilities management has been the recognition that facilities are more than this – they are 'factors of production'. As such, their importance is much greater than their capital value or even their resulting operational cost. The tendency to look only at the left-hand side of the balance sheet (cost) has distorted our view. Recent economic pressures and budget constraints have further amplified this preoccupation with cost saving. Whilst we have very sophisticated approaches to calculating this left-hand cost column, we have only rudimentary techniques for understanding or measuring those things that should appear on the right-hand side of the balance sheet – the benefits or value.

Indeed, we are only just beginning to understand how well-managed space can positively affect our ability to function as individuals and as groups. Whilst we like to legitimise our preoccupation with 'bottom line' costs, the reality is that a far greater opportunity cost is attendant on dysfunctional spaces – a cost that accrues to the organisation at large.

For most types of facilities, it is likely that the cost of getting it 'wrong' far outweighs any short-term savings in build or operational cost. Such oversights can be observed in production facilities (where buildings can inflict inefficiencies in manufacture); health facilities (where poor design can undermine staff and patient well-being); custodial facilities (where poor layout can increase the cost of security and surveillance); offices (where monotonous and inflexible surroundings can compromise creative thought); care homes (where a 'home from home' is replaced by a sterile institutional setting). The onus is on the facilities manager to identify those hidden yet vital attributes of space and services that may be overlooked in an attempt to reduce costs. This book provides a framework for making these seemingly transparent attributes more visible.

The 'dynamics of excellence' underscores the responsive role of modern facilities management. Various management principles have been put forward for managing the uncertainty that results from change. Examples include continuous improvement and more recently lean dynamics. These management approaches describe an ongoing effort to improve products, services or processes. Such approaches have been instrumental in confronting complacent attitudes regarding the achievement of excellence. Excellence instead involves continually moving forward. Whilst these techniques have proven useful in identifying 'lags' where disconnects occur between operations, decision making and information (incremental change), more searching approaches may be required in the face of the radical changes that continue to confront organisations. Issues of resilience and uncertainty (rather than variations) have emerged, involving ground-shifting change. The challenge is no longer simply to anticipate but rather to provide agility in the face of a changing landscape. The 'dynamics of excellence' embraces both the idea of ongoing improvement and adaptive resilience. As such, the argument developed within this book considers the pursuit of excellence through using all the levers available within an environment subject to radical change. Over the life of a building such changes will inevitably occur.

At a practical level the content of the book has been significantly developed to sharpen the focus on the dynamic aspects of facilities management. Chapters 2, 4, 5 and 7 are completely new and the remainder have been refashioned and augmented with new case study material. Although each chapter remains a standalone element as before, the book now does have a narrative arc: from a focus on the opportunities of being truly user-driven in Chapters 1 to 3; to management levers in Chapters 4 to 6; followed by possible tools in Chapters 7 and 8; to Chapter 9, with its concluding focus on sustaining the pursuit of excellence in facilities management.

Finally, we return to the idea of 'excellence'. It is instructive to consider the words of Aristotle:

> 'Excellence is never an accident. It is always the result of high intention, sincere effort, and intelligent execution; it represents the wise choice of many alternatives – choice, not chance, determines your destiny.'

Good facilities management never happens by chance. It is hoped that the ideas developed within this book will enable professionals to achieve a balanced consideration of all of the possibilities afforded by physical space. By so doing they will be able to make a more informed choice, cognisant of both intended and unintended consequences. As in the previous editions, the overall objective of this book is to complement and stimulate the facilities manager's own knowledge and expertise, whilst also contributing to the development of a shared knowledge base for the facilities management discipline.

Peter Barrett
Eddy Finch
June 2013

Acknowledgements

The First Edition of this book, published in 1995, was the result of a collaborative effort between a team of people from both industry and academia. Half of the resource came from industry, but the other half was provided by the Engineering and Physical Sciences Research Council and the Department of the Environment through the LINK CMR programme and this support is gratefully acknowledged. A grant was also received from the Royal Institution of Chartered Surveyors which greatly facilitated the work.

The research that led to that edition was carried out in a wide range of areas and the following industrial partners all contributed in different ways, by providing, variously, material, information, access to situations, advice and ideas:

- Barclays Property Holdings Ltd
- Chesterton International plc
- Cyril Sweett & Partners
- Ernst & Young
- Nuffield Hospitals
- The Royal Institution of Chartered Surveyors

Many organisations were involved beyond the main partners, especially in the original providing the case study material in Chapter 1. Thanks are due to these organisations, despite their anonymity. Of particular note was Dr David Owen's input to Chapter 4 of the first editor on contracting-out, which is derived substantially from his doctoral work, which was associated with the project.

Martin Sexton and Catherine Stanley, who were the research assistants on the project, carried out the great majority of the fieldwork. Principal credit goes to Martin for Chapters 6 and 7 and to Catherine for Chapters 1 and 3 (of the First Edition). Several members of staff within Salford University made helpful contributions. In particular, John Hudson's advice in the areas of briefing and IT deserves mention.

Building on this sound foundation, working with David Baldry, the 2003 Second Edition updated, supplemented and filled out the material. Dr Dilanthi Amaratunga was very helpful in providing new case study material drawn from her PhD studies at Salford University.

This Third Edition is a more radical revision, driven by a new collaboration between Professors Peter Barrett and Eddy Finch. In doing this they have drawn from a variety of relevant research projects and collaborations and would like to thank their network of colleagues for the stimulus provided.

As will be evident, many people have been involved over the years. We would like to thank them all for not only helping create this book, but also for making the process so enjoyable.

Dynamic, Strategic Facilities Management

1 Diversity and Balance in Facilities Management

1.1 Introduction

1.1.1 Scope of the chapter

The aim of this chapter is to help facilities managers take an objective view of their facilities management systems to gain a fuller appreciation of the various interactive elements. This is intended as a useful precursor to an assessment of whether the various aspects are in balance and to see if there is room for improvement. A general model and a discussion of the issues around the key dimensions are provided. This is followed by a number of case studies providing real life examples of existing facilities management organisations. The case studies do not necessarily demonstrate good practice; indeed in some cases they show how not to do it. They are intended to show the wide variety of approaches that can be employed. Any suggestions for good practice should not be followed to the letter; they are intended purely to stimulate the facilities manager into thinking about the different possibilities. No two facilities departments are likely to be identical as they will be designed to meet the needs of their parent organisations.

1.1.2 Summary of the different sections

- Section 1.1. Introduction.
- Section 1.2. A generic model is presented that shows how the elements of an ideal facilities management department would interact.
- Section 1.3. This section draws together general conclusions from the case studies in Section 1.5, suggesting where the problem areas in facilities management may lie. The section goes on to consider suggestions for good practice within facilities management.
- Section 1.4. Different models are presented allowing facilities managers to identify their organisation with a particular model. Each model is accompanied by a pointer to a particular case study in the next section, which provides a real life example(s) of that model.

Facilities Management: The Dynamics of Excellence, Third Edition. Peter Barrett and Edward Finch.
© 2014 John Wiley & Sons, Ltd. Published 2014 by John Wiley & Sons, Ltd.

- Section 1.5. Case studies are used to illustrate how different organisations operate within the different models.
- Section 1.6. Conclusions.

1.1.3   How to use this chapter

The material in this chapter can be used in a number of ways:

- It can be read sequentially.
- You may wish to go straight to the suggestions for good practice.
- You may be able to identify with a specific facilities management (FM) model and go straight to the appropriate case study.
- You may be particularly interested in a specific area, such as the structure of the facilities department and hence may wish to compare across the case studies (to make this easier each case study follows the same format).
- You may find a useful reference within the text and decide to go straight to another chapter.

1.2   Generic FM model

Facilities management is complex and involves many interactions. One of the main objectives of considering a diverse set of case studies (see later in this chapter) is to extract the key interactions at a general level. Therefore, although there are many different practices at large there are also certain regularly occurring functions that have to be addressed if facilities management is to be effective.

The generic model shown in Figure 1.1 is based on a combination of systems theory and information processing perspectives (Galbraith, 1973; Beer, 1985; Kast and Rosenzweig, 1985), linked to the practical material of the case studies. It illustrates the general range of continuing interactions that are involved in facilities management. The generic model shows how an 'ideal' facilities department would interact with the core business and the external environment. The model differentiates between strategic and operational facilities management, highlighting the need to consider the future situation, as well as the current one. In each of the following examples, the term facilities manager is referred to, but, as the case studies demonstrated, it is unlikely that any one person could be responsible for all of these areas and a facilities team is more likely, quite possibly with different people responsible for the strategic and operational areas.

The different interactions are as follows, with the numbers cross-referencing to Figure 1.1.

1.2.1   Operational facilities management

1. Interaction within the facilities department itself, between the facilities manager and the different functional units. The latter are the actual operational units of the facilities department and are likely to correspond

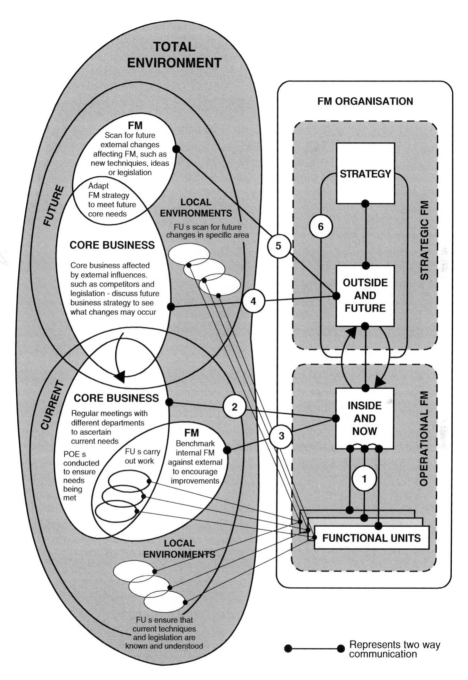

Figure 1.1   Generic model for facilities management systems.

to functions such as: maintenance, interior planning, architecture and engineering services, etc. It should be noted that the functional units can either be in-house or contracted out. With reference to this particular relationship, the facilities manager is acting in the role of coordinator, rather than implementer. The functional units are expected to carry out their duties as directed, only referring major exceptions back to the facilities manager. In this way, the latter can concentrate on the other interactions. Each of the functional units should be fully aware of current techniques and legislation relevant to their specific area. They should also scan for possible future changes and inform the facilities manager as necessary.

2. The facilities manager interacts on a regular basis with the core business to identify current facilities requirements. This could be achieved on a formal or informal basis, depending on the organisation. Audits or post-occupancy evaluations should also be conducted to ensure that these needs are actually being met and to identify areas that could be improved.

3. The facilities manager benchmarks existing internal facilities services against other facilities management organisations, so that possible areas for improvement can again be identified.

1.2.2   Strategic facilities management

4. The facilities manager interacts with the core to ascertain what future changes may occur to the business, as a response to external influences, such as competitors' plans.

5. The facilities manager will also scan for possible developments within the facilities management arena.

6. Strategy is the policy framework, which provides the context for decision-making within the facilities department. Interaction occurs between strategic and operational facilities management, the aim being to synergistically balance current operations with the needs of the future.

1.2.3   Discussion

It should be noted that the generic model should be used as a framework of the aspects a facilities manager aims to keep in mind. In reality, how the issues are handled will vary for each organisation, as will the emphasis given to particular activities. What matters is that the facilities management organisation handles each of the six interactions *appropriately* in the context of their particular circumstances. Many facilities organisations are firing on two or three cylinders, not all of the six given above. This means less power, with the dormant interactions not contributing, or in fact acting as a drag on the active interactions. For the facilities function to achieve its full potential all six interactions must be dealt with appropriately. For most organisations this will mean some scope for improvement.

Much of this book is focused on the dynamics of facilities management and, in particular, Chapters 2 and 5 focus on excellence in FM and enhancing FM performance respectively. The generic FM model stresses how facilities managers need to be responsive to the core business, but also to developments in facilities management practice and society more generally, whilst all the time having to manage the practicalities of service delivery. The remainder of this chapter turns to practical illustrations of the richness of these elements of practice across the various aspects of the model.

1.3   Illustrations of facilities management systems

1.3.1   Overview

Within the context of the above generic FM model, this section begins by summarising the findings of the case studies and first highlights the potential problem areas within facilities management systems in Section 1.3.2. The section goes on to consider suggestions for good practice within facilities management. Whilst conducting the interviews for the case studies, a standard checklist was utilised so that comparisons could readily be made across the organisations. This checklist was derived from the generic FM model set out in Section 1.2. The checklist expanded upon the following basic themes:

- facilities management structure;
- management of facilities management services;
- meeting current core business needs;
- facilities management and external influences;
- strategic facilities management.

This provides the structure for Sections 1.3.3 to 1.3.7 where suggestions for good practice are given.

1.3.2   Potential problem areas in FM

The case studies presented later in Section 1.5 provide an indication of the varied nature of facilities management. Even though eight organisations are considered, facilities management is viewed very differently by each one. In some of the organisations, for example, facilities management is expressed primarily as a maintenance function, whereas in others the scope is very much wider, including services such as catering or security. Another area where the organisations differ is whether services are provided in-house or contracted out.

Such differences are not surprising and are to be expected, as facilities departments are necessarily tailored to meet the individual needs of their particular organisation. In addition, it is still a relatively new discipline and as such is still trying to find an agreed identity. The case studies, however, do draw attention to a major issue, which is neglected by many organisations, namely the strategic relevance of facilities management. In several of the organisations, facilities management is considered to be a purely operational function. Hence, the facilities departments exist to provide a day-to-day service, not to consider how facilities could benefit the core business in the long term. In these organisations, senior management fail to comprehend that their facilities personnel possess valuable knowledge that could be utilised when making major corporate decisions.

In two of the cases, for example, the organisations had relocated. In each case the facilities department was not involved in the decision-making process and was only brought in to advise after sites had been purchased and new buildings designed. Hence, certain important factors, such as churn, were not taken into consideration and problems have occurred as a result.

The organisations that do not consider facilities management to have a strategic role are therefore neglecting a source of information that is just waiting to be utilised. However, it is not only at the strategic level that opportunities for improvement are being wasted, but also at the operational level. Communication actually within facilities departments was normally effective and the different functional units generally worked together to provide an integrated service. On the other hand, communication outside of the department, i.e. with the rest of the organization, was often ineffective, as the facilities department waited to receive instructions rather than actively asking their users what they required.

The preceding analysis indicates that there is often room for improvement within the facilities management field and so the following sections make suggestions on how these could be achieved. Even though the case studies highlight problems, they also provide many examples of well-designed facilities management systems and so the proposals can be regarded as a synthesis of good practice, as demonstrated by the case study organisations. It should be remembered, however, that all organisations are different and not all of the proposals will be applicable to every organisation.

1.3.3 Facilities management structure

The facility management models and the case studies show that there are various ways to organise the facilities department; basically there is no one method that will guarantee success. Bearing that in mind, the following points should be taken into consideration when organising a facilities department. The size of the organisation is the starting point for deciding how any facilities department is to be structured. Different sized organisations will require different staffing levels. If an organisation is quite small and located in just one building, for example, there is probably no need for a full-time facilities manager, as the amount of facilities work undertaken will be minimal. At the other end of the scale, a large organisation may need a correspondingly large facilities department.

Location is also important. If a facility department is dealing with multiple sites it will undoubtedly require a different approach to one operating on a single site. With a multiple site organisation, the facilities manager will have to decide whether services are to be provided on a centralised or decentralised basis. It is likely that a certain amount of autonomy must be granted to each site to make

Table 1.1   Typical facilities management activities.

Facility planning	Building operations and maintenance
• Strategic space planning • Set corporate planning standards and guidelines • Identify user needs • Furniture layouts • Monitor space use • Select and control use of furniture • Define performance measures • Computer-aided facility management (CAFM)	• Run and maintain plant • Maintain building fabric • Manage and undertake adaptation • Energy management • Security • Voice and data communication • Control operating budget • Monitor performance • Supervise cleaning and decoration • Waste management and recycling
Real estate and building construction	**General/office services**
• New building design and construction management • Acquisition and disposal of sites and buildings • Negotiation and management of leases • Advice on property investment • Control of capital budgets	• Provide and manage support services • Office purchasing (stationery and equipment) • Non-building contract services (catering, travel, etc.) • Reprographic services • Housekeeping standards • Relocation • Health and Safety

everyday facility decisions or else services could grind to a halt. For example, in the case of the professional group (Case Study 4), an assistant facilities manager is located at each site to deal with day-to-day operations, leaving the head facilities manager free to address major problems.

Another major consideration for the facilities manager is what services should be provided by the facilities department. Again there is not a definitive guide as to what should be included. The case study organisations, for example, vary considerably in their choice of functions; some concentrate primarily on maintenance, whilst others include general office services. As a rough guide, any facilities department is likely to perform some of the activities listed in Table 1.1 (Thomson, 1990). However, facilities managers should not just select items from the list at random, but provide only those services that are needed by their particular organisation. Once established, facilities departments do not have to limit themselves to their original activities and so the list can be extended as necessary. It is notable that in its survey of Facilities Managers' Responsibilities 1999 the British Institute of Facilities Management (BIFM) (1999) identified 29 distinct functions that a significant number of its membership carried out.

A trend in many organisations seems to be that the *conception* of what should come within the ambit of facilities management is changing. Therefore, although, for example, an organisation may have traditionally used an architect to do major refurbishments as something separate from a maintenance orientated role for the facilities department, it may decide to put all of these activities under the facilities banner. The architect may well still do the major refurbishments, but his point of contact will be the facilities manager and the building related issues of the organisation will be more closely integrated.

A further decision to be made relating to the choice of services is whether they are to be provided in-house or contracted out. The latter has gained in popularity, but as the case studies demonstrate there are no hard and fast rules concerning what should be kept in-house and what should be contracted out. Some organisations favour a totally in-house option, while others literally contract every service possible, and then there are those that will use a combination of both. Due to the number of possibilities and issues involved, contracting out is contextualised and discussed further in Chapter 4.

The background of personnel may be another influential factor when deciding how to staff a facilities department. As facilities management is still a relatively new profession, there are a limited number of people as yet who possess qualifications in this specific field. Most facility managers, therefore, will have previously trained or worked in other areas – sometimes in related professions like surveying, but often in totally different areas like human resources. A lack of technical skills is not necessarily a problem, as the facility manager's role is to coordinate work, not implement it. Indeed, several of the case study organisations had chosen to appoint existing staff as facilities managers. The reasons put forward to support these decisions included: they had proven track records as managers and they were already familiar with the operations and culture of the organisation. These organisations complemented these existing skills by sending their facilities managers on courses to acquire the necessary basic technical knowledge. Another approach used by some of the organisations, where there were assistant or regional facilities managers, was to employ people from different disciplines who could support each other.

1.3.4 *Management of facilities management services*

A facilities manager can be responsible for the provision of many varied services, as Table 1.1 shows. A common mistake made by many facilities managers is to think that they have to be involved at every stage of the delivery process and know every last detail about what is happening, but it should be remembered that it is a facilities manager's role to coordinate or, as the name implies, manage these services. Only when facilities managers learn to manage effectively and efficiently will they be able to turn their attention towards strategic issues, which is where facilities management may really be of use to its core business. So how can facilities managers make time to consider strategic considerations?

Information overload is a major problem for many facilities managers, who find that they spend all their time attending to basic operational problems. Hence, the facilities manager should empower other members of the team to make decisions, encouraging problems to be addressed at lower levels in the hierarchy. Depending on the nature of the problem, this could mean either the functional units or assistant facilities managers. For example, in the case of the school (Case Study 2), the different functional units worked together initially to sort out problems and only approached the facilities manager (bursar) with major difficulties. In organisations with various sites, such decentralised decision making will be essential if operations are to be maintained.

A further way to ease information overload is to ensure that all of the facilities team, both in-house and contractors, know exactly what is expected of them. Thus it is often worth establishing procedures to address this issue. In larger organisations this will probably mean making use of formal work programmes, service level agreements, maintenance schedules, etc., as briefing tools. Regular meetings to discuss workloads and performance may also be useful. It should be remembered, though, that informal methods can be utilised as well and may be just as effective, particularly in small firms. In the school (Case Study 2), for example, the bursar held formal weekly meetings to discuss workloads, but he also checked on progress while he walked around the school attending to other duties. Consequently problems could be sorted out on the spot, rather than waiting for the next meeting.

Investment in information technology may be another way to make information processing easier. This is becoming a popular option and there is an ever increasing number of specialist facilities management software packages appearing on the market. These packages offer a variety of different features and so the facilities manager should evaluate possible systems to ensure that they meet an organisation's particular needs. In some cases, the facilities manager may find that IT solutions are just not appropriate, as in the case of the corporate headquarters (Case Study 3), when the facilities manager found that all of the systems he reviewed were far too complex for his requirements. In order to help the facilities manager make appropriate decisions regarding this subject, the issues involved are covered in more detail in Chapter 8.

Even though facilities managers have easy access to a variety of information sources, opportunities to utilise or manipulate information are frequently wasted. Facilities managers are often responsible for a number of buildings and therefore they should perhaps consider making comparisons across buildings to identify where improvements or savings could be made. Such internal benchmarking can be used in a variety of ways. Many facilities managers, for example, will have data relating to energy consumption of their buildings. These figures could be compared to see if certain buildings were performing better, reasons for this could be established and perhaps applied elsewhere. In a similar vein, the hospital (Case Study 5) used internal benchmarking to see if it was more cost-effective to employ in-house staff or contractors to perform a specific function.

1.3.5 *Meeting current core business needs*

Even though facilities management exists to support the core business, it is often this relationship that runs into difficulties. As it is a support service, many facilities managers have taken on a reactive role, waiting for instructions before they perform any action. This often means that dialogue will only occur when problems arise. The result is that the facilities manager has to remedy the situation quickly, rather than assessing what would be the best long-term solution. It would be far better in some cases if the facilities manager had time to discuss the various implications. Such a lack of consultation is likely to result in a facilities management service that does not necessarily support the core business to the best of its capabilities. A typical example of this lack of communication would be an office move. Ideally in this situation the facilities team would consult with the users to find out how each person worked and who they needed to be located next to. However, facilities groups are rarely given enough time to do this and so the users are often moved into an impersonal office space that does not support their particular working patterns. Consequently the whole department is likely to be demoralised and productivity may be reduced.

One of the ways to improve facilities services therefore is to become more proactive, i.e. actively seek out problems and requirements before they become critical. In several of the case study organisations, this meant arranging regular meetings to discuss the services provided by the facilities management group. In Case Study 3, for example, formal meetings are held every two months, which are attended by the facilities team and representatives from each department, who will have been briefed prior to the meeting. Furthermore, proactive behaviour could include addressing emerging areas of significant activity, such as environmental management and business continuity planning.

In some organisations, staff are not the only people who will be on the receiving end of facilities services. In the private healthcare group (Case Study 5), for example, facilities management efforts are directed towards making a patient's stay as pleasant as possible. In a situation where the users are not part of the organisation, it is not always possible or sensible to try to ascertain what they think of facilities management services. Therefore, facilities managers should try to target people that will provide them with useful information. In the case of the hospitals, it can be argued that it makes more sense to discuss the provision of services with people who are there full time and who can speak on behalf of the patients, namely the nurses and consultants. Even though meetings are a useful way of gauging satisfaction with facilities services, there is generally not time to discuss things in great detail and only certain people's views will be represented. Facilities managers should therefore consider developing an audit system that seeks to improve services through feedback.

A variety of techniques have been developed to allow facilities managers to do this and can be grouped together under the title of post-occupancy evaluation (POE). At its most basic level, POE is a formal assessment of a building by its occupants after it has been completed or occupied, to identify areas that do not meet users' requirements. However, despite its title, POE is also useful when planning new facilities or altering existing ones, as data generated during an evaluation can be used in the briefing process for a new project. Due to its flexibility, POE is a tool that will be useful at various times for many facilities managers and hence is covered in detail in Chapter 3.

1.3.6 *Facilities management and external influences*

Facilities management is a very wide field and consequently a continually changing one. New legislation and new techniques are appearing all the time and it would be virtually impossible for one person to keep track of all the different changes. Therefore, the facilities manager needs to employ certain methods to make this information processing task easier.

Firstly, the facilities manager should utilise the expertise that already exists within the department. The facilities manager's role is that of coordinator; therefore, each of the functional units should ideally ensure that it is fully aware of developments within its own area of expertise and report any significant changes to the facilities manager. This should apply to both in-house personnel and contractors. The facilities manager will often have to take positive action to enable the functional units to acquire this knowledge. For example, one of the case study organisations sends its maintenance technicians on regular courses to guarantee that they are fully aware of the latest techniques and legislation.

Secondly, another way for the facilities manager to keep abreast of changes is to make use of existing external contacts. Facilities managers have to deal constantly with many different specialists as part of their work, such as insurance firms, fire officers, building control, etc. Therefore, it makes sense to maintain good communications with these people so that they can advise on new developments in their areas. In the case of the school (Case Study 2), for example, the facilities manager has established a strong working relationship with the local fire service, which carries out frequent fire inspections to check that current standards are met and also advises on potential changes. In this way the school can plan refurbishment work with the new changes in mind.

Thirdly, facilities managers may also find it helpful to make contact with other local businesses and exchange ideas. One of the case study organisations (Case Study 3), for example, is located in a business park and so the facilities manager attends residents' meetings to discuss mutual concerns. As a result of these meetings, the facilities managers have established a local benchmarking group, whereby they visit each other's buildings to study at first hand how different organisations operate. With benchmarking the number of possibilities for gaining information is almost limitless and depends purely on the nature of the relationship between the participants. Benchmarking could be used to compare processes, services, performance of plant, etc.

Finally, the facilities manager can take advantage of the growing number of specialist information sources dedicated to facilities management. These include:

- professional associations, such as BIFM (British Institute of Facilities Management);
- books;
- periodicals;
- conferences;

- short courses;
- postgraduate degree courses;
- collaborative research projects (joint academic and industry).

1.3.7 Strategic facilities management

Some of the organisations had come to realise that facilities had an important role to play in strategic planning. The private healthcare group (Case Study 5), for example, had recognised that they were not only judged on their medical care but also on the physical state of their hospitals and ancillary services, such as catering, both of which fell under the facilities umbrella. Therefore, in order to remain competitive, an appropriate facilities strategy was essential. Indeed, facilities issues have become such a major concern in this organisation that the facilities manager has been appointed to the board, so that he is involved fully in strategic decision making.

It should not be assumed, however, that only larger organisations can benefit from strategic facilities management. It can also play an important role in smaller organisations, as was demonstrated by the independent school (Case Study 2). When one of their buildings suddenly became vacant, the facilities manager took the opportunity to devise a comprehensive facilities strategy. This in turn led to an improved layout for the whole school, incorporating a number of new well-equipped facilities. Hence, the school has gained a certain amount of competitive advantage as it can now offer additional subject areas.

1.4 Facilities management models

1.4.1 Context

Experience has demonstrated that facilities management departments vary considerably from one organisation to another. This is due to the fact that they have developed in response to the particular needs of their organisation. Despite these differences most facilities departments generally fall into one of five categories (Cotts, 1990):

- office manager;
- single site;
- localised site;
- multiple site;
- international.

These models focus primarily on location, and therefore indirectly size, but this is only one method of classifying facilities departments.

Facilities managers may want to try to identify their organisation with a particular type and then go on to read the associated case studies to see if there are any similarities. It should be noted that the models are not core business specific and so even if the case study is of a different organisational type, the facilities manager should still find some similarities.

1.4.2 Office manager model

In this model, facilities management is not usually a distinct function within the organization; instead it is often undertaken by someone as part of their general duties, such as the office manager. There are two possible reasons for this. In the first instance, the organisation is located in just one building, which is too small to warrant a separate facilities department/manager. Alternatively, the organisation may be located in a leased building and hence will not want to devote personnel resources for facilities management in a building that they have no real control over. Any necessary facilities work is likely to be undertaken by consultants or contractors on a needs basis. Hence, facilities management in this case is primarily conducted through the administration of service contracts and leases. Any facilities related activities are likely to be reactive, rather than proactive.

Case Study 1: Small manufacturing firm

The firm specialises in innovative equipment for the healthcare industry. The organisation is located in a factory unit that was built three years ago, specifically for them. As the building is so new and relatively small, facilities management is not a distinct function and hence facilities related activities are undertaken by the office manager, as part of his general duties.

1.4.3 Single site model

This model applies to organisations that are large enough to have a separate facilities department, but are located at just one site. Consequently, it is the most straightforward example of a full-service facilities organisation. In this category, organisations tend to own the buildings that they occupy and therefore are prepared to spend more time and money on them than in the previous example, and hence the establishment of a separate department to deal solely with facilities issues. These organisations will probably use a combination of in-house and contracted services, but the balance of the two will vary from organisation to organisation.

Case Study 2: Independent day school

The school provides co-education for over 1000 pupils. It consists of buildings of various ages, with some over 100 years old, grouped together on one site. Facilities management is the responsibility of the bursar, assisted by a small facilities team who deal mainly with the maintenance of the buildings and associated grounds.

Case Study 3: Commercial organisation's headquarters

The headquarters provide office accommodation for 600 people in a building located in a business park. A separate facilities department has been established to deal just with this site due to the large number of people housed there. Three full-time facilities staff are responsible for the coordination of a number of contracted functions.

1.4.4   Localised sites model

This model is generally applicable to organisations that have buildings on more than one site, most often within the same metropolitan area. Typical examples would be an organisational headquarters with branches located nearby or a university with several sites. However, the same principles could easily apply to an organisation with just a couple of buildings in different parts of the country.

The idea of decentralisation comes into play with this model, as it is probable that simple operational decisions can be made at the lesser sites, while problems are passed back to the headquarters. Complete decentralisation is unlikely in this case due to economic constraints. This model will probably have a combination of in-house personnel and consultants/contractors in order to deal with the time–distance factors involved. The more decentralised the organisation, the more probable it is that external personnel will be used. In all cases, however, the headquarters will provide policy, overviews, budget control and technical assistance.

Case Study 4: Professional group

The group operates from two locations: an old headquarters building in London and a regional office in a new business park located at some distance from the capital. The facilities team consists of a facilities manager based in London, who is concerned with general facilities policies, and two assistant facilities managers, one in each building, who are responsible for daily operations.

1.4.5   Multiple sites model

This model is applicable to large organisations that operate across widely separated geographic regions, probably nationally. In this model the major headquarters is primarily concerned with policy and providing guidance to sub-ordinate regional headquarters. The principal functions are allocating resources, planning (both tactical and strategic), real estate acquisition and disposal, policy and standard setting, technical assistance, macro-level space planning and management, project management and overview. Operational issues tend to be de-emphasised, apart from when they relate to the major headquarters itself, and are dealt with at regional level.

Case Study 5: Private healthcare group

The group has 32 hospitals located around the country. Facilities management exists at four levels: board, corporate, regional and hospital. The facilities director assisted by the corporate level provides guidance on general facilities policies. The latter also oversee new/refurbishment work. The regional level undertakes a coordinating role and the hospital level covers daily operations.

Case Study 6: Historic property group

The group acts as the managing agent for over 350 historic properties, so it can actually be considered as a professional facilities management firm. Many of the properties are open to the public. The group has a three-tier management system: regional, local and site-based. At present most services are provided in-house, but there are plans for privatisation in some areas.

1.4.6 International model

This model is very similar to the previous example, but applies to large international organisations rather than national ones. The facilities department located at the headquarters will again act as policy maker and resource allocator, whilst the regional/national offices will be primarily self-managing and responsible for operational activities. It should be remembered, however, that allowances will have to be made to accommodate possible differences between the countries involved, such as legislation and language.

1.4.7 Public sector model

The following public sector case studies emphasise situations where policy decisions and processes are influenced by powerful factors which are often of a non-financial nature but relate to standards of public service provision, public probity and accountability, and the need to meet the expectations of a diverse and influential collection of stakeholder interests. Organisational change is an endemic feature of contemporary public sector organisations as they seek to respond to a broad range of dynamic environmental forces.

Case Study 7: NHS healthcare trust

This NHS trust situated in the North West of England provides a wide range of services to a multi-cultural and diverse population. The trust recognised that the cultural and economic diversification of its patients and visitors required services that are timely and sensitive in approach, ensuring the delivery of appropriate local and regional healthcare services. This organisation has been deemed as representing 'best practice' amongst the different trusts investigated during recently concluded research, on account of the mechanisms provided for managing performance measurement in facilities management.

Case Study 8: University

This case is a university based within the heart of the modern centre of a major city in the North East of England. This university was recognised as the most successful UK institution for widening participation in higher education and has received substantial government funding to enhance this work. There was a large estate and its efficient operation and maintenance were very challenging due to the scattered nature and varied age and suitability of premises; this was the major function of its facilities management department.

1.5 Case Studies

1.5.1 Overview

The following case studies provide real life examples of a number of facilities management organisations. The case studies demonstrate that facilities managers employ a variety of different approaches to similar situations and problems. However, the case studies do not necessarily demonstrate good practice; indeed in some cases they positively show how things should not be done.

Whilst conducting the interviews for the case studies, a standard checklist was utilised so that comparisons could readily be made across the organisations. This checklist was derived from the generic FM model set out in Section 1.2. The checklist expanded upon the following basic themes:

- facilities management structure;
- management of facilities management services;
- meeting current core business needs;
- facilities management and external influences;
- strategic facilities management.

Consequently, to enable comparisons to be made across the case studies the same headings are used throughout.

1.5.2 Office manager example

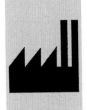

Case Study 1: Small manufacturing firm

The firm specialises in innovative equipment for the healthcare industry. The organisation is located in a factory unit that was built three years ago, specifically for them. As the building is so new and relatively small, facilities management is not a distinct function and hence facilities related activities are undertaken by the office manager, as part of his general duties.

Background

The organisation is a small manufacturing firm that produces specialist technical equipment, particularly for use in healthcare. The firm is actually part of a larger group with five other sites located around the same city. The group as a whole is very customer driven and new products are developed in line with the latest requirements. Hence, this particular building was erected three years ago to house a new product. Even though the firm is part of a larger group it actually operates as an autonomous unit and so for the purposes of this study can be regarded as a separate entity. The building itself is a typical 'shed type' factory unit, with the majority of the floorspace dedicated to manufacturing and the remainder taken up mainly by offices.

Facilities management structure

The scope of facilities management is naturally limited within small organisations, particularly in non-office situations like this factory, where changes to layouts etc. are minimal. In this case therefore facilities management corresponds basically to maintenance considerations. A full-time facilities manager would be inappropriate and so facilities linked operations form just part of the office manager's general duties. Similarly, there is no point in employing in-house staff to actually carry out occasional maintenance activities and so the majority of these functions are contracted out, including HVAC and plumbing. However, the organisation employs a full-time electrical engineer, who is responsible for both the building and the production machinery. Cleaning is also retained in-house as it is necessary every day.

Management of facilities management services

In this particular organisation, the role of the facilities manager (office manager) is relatively straightforward: basically it is his responsibility to ensure that all maintenance work is carried out as necessary. In the case of the contractors this involves checking that they carry out regular servicing as laid down in their contracts. In addition, if any failures or problems occur, it is his duty to arrange for them to be corrected.

Meeting current core business needs

The fact that this is a manufacturing firm means that it is likely to have different expectations, as far as facilities management is concerned, to organisations whose work is predominantly office based. Most of the machinery is in fixed locations and does not need to be moved very often and so most facilities requests will concern technical failures, such as the ventilation system not working. In this particular situation formal procedures for assessing current core business needs are not really necessary. If problems occur people will go directly to the office manager to get things sorted out. However, there are certain areas within the building that are not related to production, which the facilities manager feels could be improved, such as the rest areas and the cafeteria; hence he is planning to ask the staff if they are satisfied with these areas.

Facilities management and external influences

Even small organisations are not immune from external influences and this firm is no exception. The organisation's insurance company recently conducted a risk assessment of the firm and as a consequence a new condition of their insurance was that a disaster recovery plan was drawn up. The office manager therefore had to devise a contingency plan, stating what the firm would do in case of power failures or fire. The office manager is also responsible for health and safety within the organisation. However, the subject actually gets very little attention and he relies mainly on junk mail to keep him informed.

Strategic facilities management

In an organisation of this size, certain strategic facilities considerations are likely to be automatically linked to the core business strategy. For example, the introduction of a major new product line in this organisation normally involves building a new factory. Also in this case the hierarchy is very flat and so the facilities manager is involved naturally in the strategic decision-making process.

Comment

At the moment facilities management as a whole is actually quite a minor consideration in this organisation. However, small organisations such as this one will probably have to abandon such a laissez-faire attitude in the future as further legislation relating to facilities and workers' environments is introduced. As this occurs, organisations will be forced to take a more professional attitude towards facilities management and actually plan how they are going to deal with these new requirements, without totally disrupting the core business.

1.5.3 Single site example 1

Case Study 2: Independent day school

The school provides co-education for over 1000 pupils. The school consists of buildings of various ages, with some over 100 years old, grouped together on one site. Facilities management is the responsibility of the bursar, assisted by a small facilities team who deal mainly with the maintenance of the buildings and associated grounds.

Background

The focus of this case study is an independent co-educational day school. The population of the school comprises approximately 200 junior pupils, 900 seniors, 80 teaching staff and 30 administrative staff. All of the school's facilities are located on one site and there are a number of different buildings of various ages, with some dating back to the nineteenth century. The school is not large enough to warrant a totally separate facilities department and so facilities are the responsibility of the bursar. He is assisted by a small in-house team who are solely employed to attend to facilities issues.

Facilities management structure
In this school, all non-educational services are controlled by the bursar and are divided into two main areas: facilities and general office services. The facilities section is directed primarily at building maintenance and includes the following functional units, which are retained in-house due to the constant demand for these services:

- engineering: mechanical and minor electrical work (one engineer);
- groundwork: upkeep of grounds (head groundsman and four assistants);
- joinery: minor joinery work (head joiner and assistant);
- caretaking: security, movement of furniture (head caretaker and two assistants).

Other facilities related services are contracted out due to their specialist nature or fluctuating demand, such as cleaning and major building work.

Management of facilities management services: in-house services
Each of the in-house functional units is responsible for carrying out work in its own area of expertise. Once a week the bursar meets separately with the heads of each of the different units to discuss formally any outstanding work and new requirements. However, the bursar probably sees the units almost every day on an informal basis. In addition, all four functional units are encouraged to work very much as a team and are in constant communication, sorting out any minor discrepancies between them. The first three functional units work to a maintenance/refurbishment schedule, which is reviewed half-termly. However, where repair work is not urgent the school tries to accommodate work during the holidays to minimise disruption.

Management of facilities management services: contracted services
With contracted services that need to be carried out on a regular basis, such as cleaning, the contractor works to a detailed specification compiled by the bursar. Regular checks are carried out to ensure that the work is consistent with these requirements. If the contractor fails to comply then the bursar can implement the three month exit clause built into the contract.

While minor building work is dealt with in-house, major projects are contracted out. The school engages the same architect for all major works as a good working relationship has been established over recent years. Thus the school finds it easier to brief for new projects as the architect is already familiar with how the school operates. In addition, the architect is assured of an almost constant stream of work, as an increasing number of major projects are required each year in order to satisfy changing educational requirements. Major electrical work is dealt with in a similar manner, as a local electrician is permanently on call, who is also engaged during major building projects.

Meeting current core business needs

As with many organisations, there are a number of different factions at the school whose opinions have to be taken into consideration when the school is assessing current core business needs. In this case, such groups include:

- the board of governors;
- the headmaster;
- the staff;
- the parents.

To ensure that the facilities department is meeting the requirements of these groups, a series of briefing and feedback procedures have been established.

Once a term, prior to the main governors' meeting, two subcommittee meetings are held: a finance meeting and a facilities meeting. Each meeting is attended by five governors, the headmaster and the bursar. At the facilities meeting various building issues are discussed, ranging from proposed classroom alterations to the progress of ongoing building work. The bursar takes the minutes, which are distributed to the governors prior to the main meeting, so that topics causing concern can be raised and problems resolved.

The headmaster likes to keep acquainted with facilities developments through informal discussions with the bursar. However, at present this is very difficult to arrange as other school matters are taking precedence. In order to rectify this, the headmaster and the bursar are planning to formalise arrangements and hold weekly meetings specifically to discuss facilities issues.

Once a week a staff meeting is held, which is attended by the headmaster, the bursar, the administration staff and the teachers. People are free to raise any issue relating to the school that they feel needs to be discussed, including facilities topics. However, staff do not have to wait for this meeting to report problems, as people are encouraged to contact the bursar direct to request that certain tasks are undertaken. By utilising both formal and informal methods, the bursar is thus able to obtain constant feedback on facilities issues.

As parents pay for their children to attend this particular school, the headmaster feels it is important to keep them up to date on school developments. Parents are sent a newsletter once a term, which keeps them informed of all school activities, including any plans for new buildings or refurbishments. On major issues parents are given the opportunity to voice their opinions. Recently parents were sent a questionnaire that presented them with a number of possible options for future school developments. On this occasion they voted for the school to build a new information technology (IT) language building, as this was seen as a major selling point for the school. A parent association also exists, which meets once a month. This is attended by the headmaster who will refer any facilities problems to the bursar as necessary.

Facilities management and external influences

The bursar is responsible for ensuring that all legislation relating to facilities is complied with. This is obviously quite an undertaking for one person and so the bursar has to rely, to a certain extent, on other people to keep him informed of changes. The contracted caterers, for example, are responsible for ensuring that they comply with all the relevant current legislation and inform the bursar of new developments. Perhaps more interestingly, the bursar also makes use of external contacts who are not working for the school. He has developed a strong working relationship with the local fire service, who carry out fire inspections in the course of alterations to check that current standards are met and who also advise informally on potential changes in the law. In this way the bursar can plan refurbishment work with the new changes in mind.

Sometimes facilities managers working in small organisations may not be aware of new developments in the facilities management field. In order to ensure that this is not the case, the bursar has joined the British Institute of Facilities Management. Through this membership he receives literature on new services and ideas. He also attends local meetings of the group, where he has the chance to discuss different approaches and problems with other facilities managers who can identify with his situation.

Strategic facilities management

Until last year there had been no real link between facilities considerations and the core business strategy. Facilities management was seen very much in terms of maintenance and daily operations. As long as the facilities were maintained to a high standard and problems dealt with as they arose, the school could concentrate on its core business of education and consequently saw no real need for a facilities strategy; however, the school has been forced to rethink the importance of its facilities due to a number of external forces.

Firstly, over the last few years the educational establishment has undergone major changes with the introduction of the national curriculum. The school has been forced to consider how it will meet the new requirements that have been imposed on it. For example, new subject areas have been introduced which the school now has to offer. This will have both physical and financial implications, as additional space will need to be found to accommodate the new subjects and new teachers will also be required. Therefore, the school had to decide whether it could utilise existing classrooms or build new ones.

Secondly, the school is independently run and so it has to attract pupils in order to maintain an income. Increasing competition in private education means that the school now has to try harder to get new pupils. As a consequence, the school not only has to provide a high standard of education but has to back this up with modern first class teaching facilities. The school had to decide whether this could be achieved by upgrading some of the buildings or whether new ones would be necessary.

Thirdly, the school had always offered a boarding option in addition to daily attendance, but last year the number of boarders at the school dropped so dramatically that this mode of attendance was dropped. Consequently, the boarders' building became vacant and so the school suddenly had extra space to play with. This had two major implications for the school: firstly, refurbishment work could be carried out more easily as the space could be used as temporary accommodation; secondly, taking account of the extra space, the layout of the school could perhaps be rearranged so that departments were located more logically and new specialist areas created.

The combination of all these three factors meant that the school had no choice but to consider how its facilities should be used in the future. Hence, the school has since developed a core business strategy, which covers the next ten years and a corresponding facilities programme. The latter was achieved in the following manner. Initially a building condition survey was carried out of the whole school. This identified which buildings could be refurbished and which should perhaps be demolished. From this the bursar produced a facilities strategy that established ideally in which order the buildings should be refurbished. At the same time the bursar worked with other staff members to produce a new layout for the school that took the latter into consideration, as well as the core business strategy. Finally, a ten year programme was drawn up that details exactly what building and maintenance work needs to be carried out so that the new layout can be achieved.

Comment

Even though the school is too small to warrant a full-time facilities manager, the facilities systems are actually quite well developed. Communication is the key to efficient and effective services in this organisation. Both formal and informal communication networks are utilised to ensure not only that the work is done but that the user's needs are met. Strong links have also been established with external consultants so that the school is fully aware of new developments relating to facilities management.

As regards strategic facilities management, this was really seen as an unnecessary complication until external pressures actually forced the school to consider how their buildings contributed to the overall success of the organisation. Now that the school has studied their future options they have come to realise that the rationalisation of their buildings may help them to gain a competitive edge and thus facilities management may be about more than just maintenance. Consequently facilities implications are likely to be assessed when considering core business strategies from now on.

1.5.4   *Single site example 2*

Case Study 3: Commercial organisation's headquarters

The headquarters provide office accommodation for 600 people in a building located in a business park. A separate facilities department has been established to deal just with this site due to the large number of people housed there. Three full-time facilities staff are responsible for the coordination of a number of contracted functions.

Background

This is a large commercial organisation. A subsidiary group has been established which is responsible for property related issues. The latter has three main divisions, with the following responsibilities:

- property: investment, development, acquisition/disposals;
- professional services: managing agents, refurbishments, valuations, rent reviews;
- facilities: house management in headquarters buildings.

The facilities department therefore is concerned solely with the day-to-day running of the headquarters, while the property division is responsible for strategic planning.

Facilities management structure

Facilities management within this particular organisation covers a wide range of activities, but the organisation has chosen to contract out the majority of these functions and so the facilities department consists of an in-house management team of only three people who ensure that the various contractors complete their duties. Broadly speaking, these contracted-out functions can be divided into two areas: building/maintenance and general/office services (Table 1.2).

Management of facilities management services

The in-house facilities management team is responsible for ensuring that all of the above activities are completed as agreed in the relevant contracts. The contracts vary according to the activity; some people work for a fixed number of hours per week, whereas others are on time plus materials. All of the outsourced functions provide quotations before they are awarded the contract to ensure that they are competitive. These quotations are compared regularly against the competition to check that the organisation is still getting value for money. As there are so many outsourced functions, procedures for briefing the contractors have to be tightly controlled and so formal work orders are issued for every job. Once work has been completed it is checked off in an order book and contractors can then issue invoices.

Table 1.2   Contracted-out functions.

Building / maintenance	General / office services
• Mechanical and electrical (M&E) maintenance • Grounds maintenance • Building contractors • Furniture alterations • Office moves and changes • Cleaning • Day janitorial service • Waste disposal	• Security • Office administration • Reception • Telephones and switchboard • PABX • Mail room • Newspapers • Taxis • Press control • Catering

Even though the various outsourced personnel come from different companies they are encouraged to see themselves as part of the facilities team. Consequently, there is a good working relationship between the different outsourced people who work on the site. If contractors notice a problem that is not part of their job, they will then point it out to the people concerned.

In this organisation facilities management covers a wide range of activities, coordinated by only three in-house managers who, consequently, are sometimes overloaded. Hence, the facilities manager considered purchasing a facilities management software package to ease the situation. Unfortunately an analysis of the various packages available proved that they were all probably too complex for this particular site and provided a lot of features that would not be used. So the department will have to continue to rely on paper methods, although the department does make use of information technology in other ways. For example, to assist with energy management the organisation has installed a lighting control system. All of the lights are switched off automatically in the evening to save energy, but they can be turned on via the telephone if necessary by cleaners or people working late. Also an internal phone directory is on the network, so that the receptionists are not overloaded.

Meeting current core business needs

Facilities management activities in this organisation are very much user driven. Minor operational problems are dealt with on an informal basis, with users contacting the facilities team directly, who will attend to the problem as quickly as possible. However, the facilities team felt that it was important to actually seek out feedback from their users, rather than waiting for the users to approach them. Consequently more formal procedures have also been established and a facilities meeting is held every two months.

This meeting is attended by two members of the facilities management team and one representative from each department, normally an administration or finance manager. Each departmental representative acts as spokesperson voicing the concerns of people in their department, who will have been consulted previous to the meeting. The representatives are also updated on what has happened as a consequence of the last meeting and any plans for the near future. Recently the issue of staff working late and the resulting security/cost problems have been a major discussion point at these meetings. Therefore, the facilities team has had to investigate how high levels of security can be maintained, whilst allowing staff to leave the building whenever they want to.

Facilities management and external influences

As the headquarters building is located in a new business park, there are certain issues that may affect all of the organisations in the park. Hence, a residents' group has been formed that meets regularly to discuss mutual concerns. The issues tend to relate to physical problems, such as new building work or a lack of car parking and so the facilities team are the most suitable people to attend the meeting.

The meetings are also useful as they allow the facilities team to make contact with other facilities managers. Thus a local network has been set up where facilities managers visit each other's buildings to study different facilities management approaches at first hand.

Strategic facilities management

As stated earlier, facilities management in this organisation is viewed as a purely operational function providing daily services that ensure the smooth running of the headquarters buildings. The organisation believes that the facilities department has no real contribution to make to strategic planning as the property division deals with this. This is probably true as far as the organisation's other buildings go, but the facilities department is not even consulted about decisions that affect the headquarters site. When decisions are made by the core business, the facilities department is often the last to be informed, even though it will have to implement the changes. For example, when the core business decides to move a department to a smaller area within a building, it is up to the facilities department to somehow fit in all of the workstations. This means that the facilities department is forced into a reactive way of working, which makes it hard to find the time to discover what the users really want.

Such a lack of communication means that the core business sometimes makes decisions without considering all of the relevant factors. A major example of this occurred when the organisation actually relocated to its present site. The organisation decided to move out of London and build a new headquarters. However, the facilities team was not brought in until the design had been practically finalised. It transpired that no allowance had been made for 'churn' (the physical reconfiguration of offices and workstations) in the design and so future changes would not be easily implemented. No doubt if the facilities department had been involved earlier it would have drawn attention to this fact.

Even though the facilities department is not likely to become involved with strategic planning, there are plans to extend its services. Over the past few years it has gained a lot of experience in managing daily operations. Hence, the team intends to offer its services in a consultancy capacity to other organisations. This suggestion has been accepted in principle at board level and so the department now has to develop the idea and identify potential customers.

Comment

The organisation is unusual in that it contracts out the majority of its facilities management services, but the arrangement appears to work extremely well in this case. Perhaps the most noticeable effect of this approach is that the facilities systems tend to be much more formal and structured than in the other case study organisation. This is because the contracted staff are not necessarily always on site and so workloads have to be planned carefully to make the best use of people's time.

As the examples above demonstrate, the facilities department has not been involved in strategic decision making in the past. Although strategic planning falls within the realm of the property division, it would actually make sense if the transfer of information was improved between the two departments, because the facilities team actually possesses valuable knowledge that is not being utilised at present.

1.5.5 Localised sites example

Case Study 4: Professional group

The group operates from two locations: an old headquarters building in London and a regional office in a new business park located at some distance from the capital. The facilities team consists of a facilities manager based in London, who is concerned with general facilities policies, and two assistant facilities managers, one in each building, who are responsible for daily operations.

Background
The focus of this study is a professional organisation, operating from two buildings: an old headquarters building located in London and a regional office in a new business park. The London headquarters houses a conference centre, a retail outlet, a library, a members' club, a restaurant and a number of offices. The regional office is used in the main for administrative purposes and is therefore mostly office space, with additional conference facilities.

Facilities management structure
Facilities management in this organisation covers a wide range of services and hence a separate facilities department is necessary. A core management team comprises a head facilities manager based at the London headquarters and two assistant facilities managers, one at each site. The head facilities manager is concerned with general policies and major issues/problems, whilst the assistants are responsible for supervising day-to-day operations. The different facilities services that are provided are listed in Table 1.3.

These services are carried out by a combination of in-house and contracted personnel. As a general guide, in-house staff undertake activities that are required

Table 1.3   Different facilities services.

Premises	Office services	Central services
• Building maintenance • Decoration works • Building sub-contractors • Telecommunications • Security • Porterage • Safety • Cleaning	• Mailing • Stationery • Photocopying • Vehicle fleet • Printing • Courier for regional office	• Catering • Conference bookings • Insurance • Archival filing

on a constant basis, such as cleaning and building maintenance. Furthermore, the facilities manager also believes it is sensible to use in-house employees for tasks that need to be tightly controlled. This particularly applies to functions that interface with the public, such as reception and the switchboard, as poor service in these areas will reflect badly on the rest of the organisation.

Not all of the above services are needed every day and hence certain functions are contracted out, as there is not enough work to justify permanent members of staff. Plant maintenance, for example, is carried out by a contractor who comes once a week to check that the boilers etc. are functioning correctly. Thus any problems can hopefully be identified before any real damage occurs; however, in the case of an emergency the contractor can be called out at any time. A second group of functions are contracted out due to their specialist nature. Catering falls into this category, as it is subject to particularly stringent health regulations.

Management of facilities management services

One of the main difficulties faced by the facilities department in this particular organisation is how to provide services for two separate locations. Owing to the distances involved, it has been necessary to duplicate some operations for each site, such as cleaning and porterage. Even though these services could be managed from the London office, it has been decided that on-site management allows problems to be addressed more effectively. Therefore, an assistant facilities manager is located at each site, who is responsible for the management of daily facilities activities and who can make simple operational decisions without referring to the head facilities manager. However, in this organisation, complete decentralisation would be an unnecessary expense, as certain activities do not need to be duplicated at both sites. Conference bookings for both locations, for example, are dealt with in London and all facilities budgeting is also centrally controlled.

Communication within the facilities department is also a potential problem area for this organisation, as the facilities team has been split up so that both sites are covered. In a single site situation, members of the facilities team would probably see each other every day and so communicate any developments on an informal basis. In this case it is important to ensure that the assistant facilities manager located in the regional office does not feel isolated from the rest of the facilities organisation. Consequently, formalised communication channels have been established, which keep the assistant facilities manager in touch with the rest of the facilities team. Thus the head facilities manager visits the regional office once every three weeks to check that things are running smoothly. In the meantime, the assistant facilities managers are encouraged to communicate frequently and sort out any problems between them. Technology also plays a part in maintaining communications; for example, an electronic mailing system is used to inform the regional office of advance conference bookings. Finally, a daily dedicated courier service runs between the two sites, which can be used as necessary by the facilities department.

Effective communication is also an important consideration for each individual site, as well as between sites. Therefore, two-way radios have been introduced so that certain functions in each building have constant access to the assistant facilities managers. Problems can then be addressed immediately even if the facilities manager is not at his desk.

The facilities manager also believes that a thorough understanding of the way the organisation operates and its culture is essential for the provision of high quality facilities services. Consequently, he prefers to appoint new facilities members from within the organisation where possible. Thus the assistant facilities manager in the regional office used to be the head porter. Similarly, the London assistant was formerly in the post room, but had a very good knowledge of computers and telecommunications and so he was promoted.

Meeting current core business needs

Much of the London headquarters building is given over to facilities for members and other visitors, including: a conference centre, library, members' club, restaurant and shop. Indeed, at any one time there could be up to 300 visitors in the building. This means that a large percentage of the facilities management effort is directed towards these external users. As there are so many visitors and they are generally in the building for such a limited time, it has been decided that it would not be a worthwhile exercise to actively seek their opinions on facilities issues.

Unfortunately, with the main thrust of the organisation directed at its members, the organisation's employees are forced to take a back seat on most issues, including facilities. If problems occur, then internal users are free to contact the facilities manager, but their wishes are not sought on a regular basis. The head facilities manager would like to conduct internal forums to discuss facilities matters, but feels that this is unlikely to happen in the near future owing to the culture of the organisation.

Facilities management and external influences

One of the main responsibilities undertaken by the head facilities manager in this organisation is to ensure that new legislation relating to facilities issues is adhered to. Consequently, when the health and safety directives were recently introduced, the facilities manager had to consider how they were to be implemented. When he tried to discuss the directives and their implications with the senior management he could not engage any response. Unable to act on his own in such an important matter, the facilities manager formally refused to take responsibility for any health and safety issues until the senior managers agreed to discuss the situation. As a result of his action a health and safety committee was established, which has since produced a policy statement that will guide all future decisions.

Strategic facilities management

The facilities management department in this organisation was established to provide daily support services only and consequently the facilities manager is not seen as having a part to play in strategic planning for the core business. This means that the facilities team has to respond as best it can once major decisions have already been made; this is highlighted by the following examples.

Approximately five years ago the organisation had a different regional office, where the lease would not be renewed and so it was necessary to find an alternative location. However, even though it was a facilities related problem, only senior managers were involved in the selection process and they engaged external professionals to investigate the different options. It was not until an actual site for a new building had been chosen that the facilities manager or anyone else in the organisation became involved and by this time it was too late to propose alternatives. During the briefing stage of the building process, the facilities manager did become a member of the steering group that was formed to advise the architect. In reality, though, any suggestions made by the facilities manager were generally ignored, particularly over larger issues, such as size or location of rooms. The facilities team's objectives, as far as the senior managers were concerned, were not to influence the design of the building, but to purely select, supply and fit out all rooms with appropriate furniture, telecommunications, etc. However, the facilities manager feels that he could have provided useful advice had he been allowed to comment, because of his detailed understanding of how the organisation operated. This applies in particular to room sizes, as it has been established that certain rooms are definitely too small to function properly.

On a more minor scale, the facilities team recently decided that the telecommunications system in part of a building was outdated and so the telephones were upgraded throughout that section. It has since come to light that the decision had already been taken by the senior management to lease out the space, as it was surplus to their requirements. Hence, the facilities team spent money upgrading a system, just to remove it again soon afterwards.

Comment

This organisation's buildings are spread across two locations; hence the major hurdle facing the facilities team was how to provide a cost-effective and efficient service across both sites. The team has worked hard over the past few years to meet this challenge and they now have an extremely effective partially decentralised system, which ensures that daily operations run smoothly. By locating an assistant facilities manager at each site, minor problems can be dealt with promptly and so services do not grind to a halt whilst waiting for an answer from the head facilities manager.

A particular point of interest in this organisation is that the head facilities manager has appointed existing staff members as assistant facilities managers. He believes that a thorough understanding of the organisation, its requirements and its culture are more important than technical knowledge. After all, other people are employed or contracted to carry out the actual work and hence the assistant facilities managers are responsible for overall supervision and coordination. However, to complement their existing skills, the assistant facilities managers are also sent on courses to acquire a basic knowledge of the relevant technical details.

The facilities department was established purely to provide operational support and so, not surprisingly, strategic facilities management is a minor consideration in this organisation. Even when a major relocation was being planned, the facilities department was very much left in the dark. Consequently, the new regional headquarters building probably does not function as well as it might have done had the facilities department been involved at an earlier stage.

1.5.6 Multiple sites example 1

Case Study 5: Private healthcare group

The group has 32 hospitals located around the country. Facilities management exists at four levels: board, corporate, regional and hospital. The facilities director assisted by the corporate level provides guidance on general facilities policies. The latter also oversee new/refurbishment work. The regional level undertakes a coordinating role and the hospital level covers daily operations.

Background

The focus of this study is an organisation that provides healthcare services in over thirty private hospitals located around Britain, primarily in cities or larger towns. The organisation also has a separate corporate headquarters and four regional offices. It is therefore a good example of a multiple site organisation.

The organisation comprises 32 hospitals in total, each of which is managed by a general manager. The group is subdivided into four regions, with eight hospitals in each, and a regional general manager has been appointed for each region. At the corporate office, there is a chief executive and a board of directors, alongside a team of professional heads of functions covering corporate finance, operational finance, personnel, legal, nursing, paramedical services, marketing and facilities management, together with supporting staff.

The organisation operates within a three-tier general management philosophy, with strategic planning and management at corporate office level, through coordinated regional management, to general day-to-day management at hospital level.

Facilities management structure

The fact that this organisation owns over 30 buildings is reflected in the size and complexity of the facilities department. The structure of the facilities department is parallel to the structure of the organisation as a whole; hence the facilities management function is represented at the following four levels throughout the organisation: board, corporate, regional and hospital. This organisation is therefore one of a minority where facilities management is viewed important enough to have achieved representation at board level.

A larger organisation has meant that the facilities department not only has a more complex structure but that it also includes a greater number of activities.

Consequently certain functions are represented that are beyond the scope of most other facilities departments. The organisation, for example, is frequently refurbishing or adding extensions to its hospitals; therefore an in-house capital project management team forms part of the facilities group located at the corporate office.

The roles and responsibilities of the different levels are as follows.

Board level

The director of facilities has a predominantly strategic role and thus only tends to become involved with the operational side of facilities management if major problems occur. His principal functions can, therefore, be summarised as:

- representing the facilities management function to the board of governors and the board of directors;
- as one of the directors, he is responsible for advising the board of governors on the general situation of the whole group.

Corporate level

Facilities management at this level is the responsibility of the group facilities manager, assisted by a team of professional and support staff. The principal functions of the corporate facilities group can be summarised as:

- setting and policing corporate wide goals, objectives and standards for the facilities management function, in compliance with legal obligations and corporate policy;
- servicing the boards of directors and governors with group reports, statistics and policy proposals relating to facilities management;
- providing professional support and guidance to operational staff at regional and hospital level.

Regional level

The organisation is divided into four regions, each of which has its own dedicated regional facilities manager and assistant facilities manager, located at a regional office. The principal functions of the regional facilities teams relate in the main to the physical structure of the hospitals and can be summarised as:

- organising, managing and monitoring performance of directly employed staff, contractors and suppliers engaged on maintenance and project work respectively;
- advising and guiding regional general and hospital managers on all issues relating to the planning and use of their physical resources for business purposes;
- assisting hospital managers when planning for revenue and minor capital budgets and monitoring performance of facilities against allocated funds.

Hospital level

Each of the 32 hospitals has its own facilities manager, who is known as the hotel services manager. As the name suggests, this position is generally concerned with the non-medical services that are provided for the general comfort of patients. Hence, the hotel services manager is responsible for the following activities: catering, domestic services, portering, reception and maintenance. It is up to the hotel services manager to ensure that all of these activities, whether in-house or contracted out, are carried out as required. In essence, therefore, each hospital is similar to the earlier single site examples, in that they tend to focus on operational rather than strategic facilities management.

Perhaps it is worth pointing out that maintenance has a slightly different relationship with the hotel services manager than the other activities. Each hospital has its own maintenance technician, who is responsible for carrying out a planned preventative maintenance schedule, as well as any necessary breakdown maintenance. Even though the technicians have line responsibility to the hotel services managers, they also keep in close contact with their regional facilities managers. This is because the hotel service managers tend to be from non-construction backgrounds and so will not always be able to adequately address maintenance issues.

Management of facilities management services

The size and complexity of the facilities department means that it is very difficult for each of the different levels to keep track of developments elsewhere within the facilities group. Hence, certain procedures have been established to ensure regular communication across the levels. Consequently, the four regional facilities managers meet with the group facilities manager on a formal basis every two months to discuss the current state-of-play in each region. Similarly, each regional facilities manager meets regularly with the hotel services managers in that region.

Formal meetings are not the only method employed to monitor facilities developments. In the case of maintenance, for example, a manual has been developed that details a planned preventative maintenance schedule and must be followed by the maintenance technician at each hospital. Thus, when the regional facilities managers (or assistants) make programmed maintenance visits to individual hospitals, they assume an auditing role and check work against the manual. Such scheduling is necessary and works well as it is impossible for maintenance visits to occur very often. This method also allows the regional facilities managers to focus on maintenance problems, rather than routine operations.

Within the different levels themselves, informal methods of communication are used alongside the more formal methods. Hence, if a regional facilities manager has a problem, he may initially contact a colleague in another region, who may have encountered similar difficulties, rather than approaching the group facilities manager. Such teamwork and lateral communications are encouraged throughout the facilities department and so problems are often sorted out without being passed on to the next level.

Internal benchmarking is another method that is used by the facilities group to improve the services that they provide. The department has access to over 30 hospitals and four different regions; hence there are plenty of opportunities to make comparisons, as the following examples show.

One of the responsibilities of the facilities department is to ensure that the clinical sterilisers are maintained in good working order and so a specialist engineer is employed in each region. The facilities department was unable to fill the post in one region and so it was decided that this role should be contracted out. This decision was made against the wishes of senior management who wanted to maintain this function in-house, but when compared against the other regions, the contracting-out option actually proved to be more cost-effective, as the contracted engineer did not have to waste time doing fill-in jobs. Therefore, benchmarking helped the facilities group to identify an area where cost savings could be made.

Another area that is compared internally are utility costs. The group facilities manager has contracted an external organisation to compare utilities costs on a monthly basis. Hence, consumption of gas, water and electricity are constantly monitored. The results are presented on simple graphs to allow for easy comparison. The graphs are discussed at regular meetings to see if any savings can be made. However, the group facilities manager does not necessarily take the readings at face value and carefully considers the different factors that may affect the results, e.g. the size of the hospital, the age and the location. Once this system is firmly established, there are plans to bring it back in-house.

The facilities department is responsible for a large number of buildings and over the years many of the hospitals have undergone substantial refurbishment. Not surprisingly, the department was finding it increasingly difficult to maintain paper records of all the changes. When work was being carried out, the hospital and the corporate office were sometimes working on different plans. Hence an AutoCad system has now been installed and up-to-date plans are being transferred on to it, as will any future changes. Similarly, the department has started to use a database to store details of the plant in all hospitals, which will assist in formulating maintenance schedules.

Meeting current core business needs

When assessing current core business needs, the facilities department has to consider both the requirements of the staff and the patients. Obviously, due to the numbers involved, it would be impossible to consult everybody and so the facilities department has to be selective.

Within the hospitals themselves, the hospital managers are responsible for the smooth running of the hospital. Therefore, the hotel services managers work closely with the hospital managers to ascertain whether appropriate service levels are provided and sort out minor problems. The hospital manager in turn consults with the different departmental heads to see if their requirements are being met and passes this information on to the hotel services manager.

Regional facilities managers are interested in the overall facilities picture at the hospitals, rather than the daily operations. Hence, semi-formal meetings are held quarterly between the hospital managers and the regional facilities mangers to review facilities budgets and general service levels. Then once a year very formal meetings are held to set the facilities budget for each hospital. However, regional facilities managers also visit each of their hospitals about once a month to walk round and see for themselves how things are going.

As far as the patients are concerned the facilities department uses more indirect methods to see if their needs are being met and rely on the staff to inform them of any problems that may affect the patients. However, the department also obtains information from a general questionnaire that is sent out to all patients by the marketing division. The 'patient satisfaction survey' asks the patient about different aspects of the performance of its hospitals, including room details and general patient services, such as catering. Quarterly reports summarise the responses and hence problem areas can often be identified and corrected.

A large part of the facilities department's work is related to the refurbishment of the hospitals and five major projects have recently been completed. Consequently, the department has decided to carry out an evaluation exercise to see if the users are totally happy with the new facilities and, if not, to identify where improvements can be made. The aim of the exercise is to hopefully learn from past mistakes and ensure that they are not repeated in the future. Such evaluation techniques are being increasingly used by facilities managers and the subject of 'post-occupancy evaluations' is addressed in detail in Chapter 3.

Facilities management and external influences

Healthcare is a rapidly changing area, with new legislation and approaches appearing all the time. Consequently, a principal function of the corporate facilities group is to ensure that the organisation is fully aware of any new developments relating to facilities management. This means ensuring that all of the hospitals are informed of any changes and checking that they go on to comply with them.

As the organisation is a well-established name in the field of healthcare, they are often approached by other hospitals for advice. Thus the director of facilities is regularly asked to visit other establishments to discuss different approaches to facilities issues. He also finds it useful to attend conferences etc., so that he is fully aware of new ideas. Similarly, other members of the facilities department are sent on courses to ensure that their skills are kept up to date.

Strategic facilities management

The core business of this organisation is the provision of healthcare. However, as far as private healthcare is concerned, medical services are only part of the story. Patients are paying for their treatment and so they expect high quality ancillary services. This means that rooms have to be modern, private, comfortable, etc., and catering has to be of a high standard. Thus the core business has to ensure that its future strategy satisfies increasing customer expectations.

Patient expectations are not the only external pressures that may affect core business strategy. The organisation also has to consider what its competitors are planning to do, as the medical consultants will always want to practise at the best-equipped hospital in the area. Consultants are responsible for referring patients to a specific hospital and so the loss of a consultant has vast implications.

A third area affecting core strategy is the speed of change within medicine itself. Improvements in treatment are occurring continually and this means that patients are spending less and less time actually in hospital. Consequently fewer beds are necessary.

All of the above factors have facilities implications and so within this organisation facilities management has become increasingly more important – so much so, that the previous group facilities manager has been promoted to the board as director of facilities. At this level within the hierarchy, the director of facilities is involved fully in corporate decision making and therefore full use is made of his facilities knowledge and experience.

For example, the organisation is currently building a new hospital as a replacement for an existing one located nearby. Originally there were plans to refurbish the old one and just build an extension for new facilities, but the facilities director suggested that the hospital was perhaps not really worth refurbishing and it might be more cost-effective just to build a new one. Feasibility studies were carried out by the facilities department and these proved that the assumptions of the facilities director were correct. Without his early input, the organisation may have wasted a substantial amount of money.

The director of facilities has also been instrumental in a major reorganisation programme, which has led to the facilities department being in a better position to assist the core business. As a result the structure of the organisation today is somewhat different from how it was in 1988 when the current director of facilities was appointed as the estates manager. At that time the structure of the organisation meant that even relatively simple requests resulted in a complex bureaucratic process. For example, if a hospital manager wished to refurbish a small area of the hospital that included patient rooms and offices, she would have to make contact with, and coordinate the input of, over ten individuals, namely:

- regional surveyor;
- M&E services manager, who would mobilise two separate line managers;
- office service manager, who would utilise three people to cover furniture, equipment and office telephones;
- purchasing manager, who would mobilise staff to cover nurse call and fire alarm installation, piped medical gas, elevators, equipment, furniture, furnishings and telecommunications;
- hotel services manager.

Even though the respective managers would manage their own staff members, the hospital manager was still faced with the task of overall coordination, for which she was not trained and would also be distracted from her primary role of managing the hospital.

A second problem existed in that capital planning and development were carried out as a separate function altogether, with little or no communication with other corporate functions, and therefore no thought was given to the future management of the facilities that the project managers delivered. A third and final contentious issue was that the different functions reported to three separate directors, thereby placing the ultimate onus of coordination on the chief executive.

All of these factors meant that the non-core functions were not assisting the core business to the best of their ability, which meant that operational issues were often dealt with at the expense of strategic planning. Within a short space of time, the newly appointed estates manager had identified the above problems and set about trying to find a solution. His answer was to propose a complete reorganisation of the structure, so that the non-core functions were grouped together under one director to provide a totally integrated service. Over the past few years this idea has been implemented, resulting in a facilities department that now fully supports the core business.

Comment

This organisation operates from over 30 separate sites and hence is substantially larger than the previous case study examples. It is not surprising, therefore, that the facilities department is correspondingly complicated. Communications could be potentially difficult to maintain across the different levels and sites, so a series of well-defined communication networks has been established to cope with the complexity.

However, as can be seen from the narrative, facilities operations have not always been so well organised. Indeed, it has taken over five years for the organisation to develop its current facilities department and systems. Thus other organisations should not be disheartened when looking at their own facilities departments and should realise that changes or improvements cannot possibly be achieved overnight.

Even though the department is now fully reorganised, the facilities team is determined not to become complacent about the services that it provides and has adopted a policy of continuous improvement. For example, they are already benchmarking internally to see where services and costs could be improved. They are also planning to conduct a series of post-occupancy evaluations to ascertain whether users are satisfied with the newly completed hospital refurbishments.

In contrast to some of the previous organisations, facilities management in this case is actually considered to have strategic relevance. This is probably because private hospitals are not only assessed on their healthcare but also on such features as the standard of patient rooms and catering. Consequently, facilities management has become increasingly important. This is reflected in the fact that the organisation now has a director of facilities who not only advises on facilities issues but who is also fully involved in considering strategic options for the organisation as a whole.

1.5.7   *Multiple sites example 2*

Case Study 6: Historic property group

The group acts as the managing agent for over 350 historic properties, so it can actually be considered as a professional facilities management firm. Many of the properties are open to the public. The group has a three-tier management system: regional, local and site-based. At present most services are provided in-house, but there are plans for privatisation in some areas.

Background

This case study considers an organisation that is responsible for the preservation of many historic properties in this country. The organisation is actually split into two main sections: the conservation group and the historic properties group. In reality the sections operate almost independently and so for the purposes of this study, the historic properties group is regarded as the organisation under focus.

The historic properties group varies substantially from the other case study organisations, in that its core business is the management of historic buildings/sites, so in essence it can be regarded as a professional facilities management firm responsible for over 350 properties around the country. There is a complete cross-section of property types, ranging from grass mounds to castles, and everything in between. The organisation is primarily funded by a government grant (approximately 90%) and the remainder is mainly raised by opening the properties to the general public.

Facilities management structure

The organisation is divided into five regions, each of which is responsible for the management of the historic properties within its region. The regions are operated on a three-tier management system: regional, group and site-based.

- Regional. A regional director is responsible for each region and is based at a regional headquarters. He has the support of the three following groups:
  - operations officers who are responsible for the day-to-day running of the sites;
  - historic officers who commission building work and maintenance inspections;
  - design and works: architects/technical officers who provide technical back-up.
- Group. All of the regions are subdivided into smaller areas, each of which is coordinated by a group custodian. Broadly speaking, the latter act as local managers, checking that the sites are being run correctly.
- Site-based. Custodians are located at many of the sites, some are full time and others are seasonal. Their primary function is to collect admission charges

from visitors, but they also ensure that any problems at the site are reported so that they can be corrected.

In addition, the historic properties group can call on the expertise of two ancillary groups: the corporate services group, who provide administrative, legal and financial support systems and advice, and the research and professional services group, who, amongst other things, provide an in-house labour force to carry out repairs, building work, etc., as necessary.

Management of facilities management services

Each region is responsible for over 50 sites and consequently formal methods have been developed for ensuring that each site receives the attention it requires and is preserved as appropriate. The sites are divided into two types: fragile and stable. The fragile sites are formally inspected by the design and works team every year to check that they have not deteriorated and to see if any major work is necessary. The stable sites are inspected in a similar manner every three years. The inspections allow the organisation to feed this information into its rolling four year plan. Thus the organisation has a good idea of how its money will be spent over the next few years.

In addition to these major inspections, many of the larger sites have two individually tailored preventative maintenance programmes. Firstly, there is an historic programme, which details maintenance relating specifically to the building fabric. Secondly, there is a general maintenance programme, which ensures that the whole site is suitable for visitors.

As the organisation is spread over the whole country and divided into five regions, a method for achieving effective communications has been a major concern. Consequently, formal procedures have been established to ensure that regular communication takes place across the regions. Thus the regional directors meet once a fortnight in London to discuss what is happening throughout the whole of the group. To supplement these meetings, the regional directors also undertake site visits in each of the regions, so that they can actually view at first hand what has happened at specific locations.

Meeting users' needs

Unlike the previous organisations, the majority of the users in this case are not part of the organisation, but the general public. In the past the organisation has had a fragmented approach to obtaining feedback from visitors to its properties. Most frequently custodians would talk to visitors informally on site and report any relevant comments to the regional headquarters. Occasional surveys were also conducted by the marketing division. However, now that improvements have been made to a number of sites, the organisation has decided to actively seek feedback and so freepost comment sheets are to be provided, especially at the larger sites. In addition, if visitors have a complaint about a site, a formal complaints procedure has been established, details of which can be found in a new customers' charter.

Facilities management and external influences

Many of the sites that are managed by the group are open to the public, so the organisation is responsible for ensuring that the appropriate safety standards are met. However, it would be virtually impossible for each site to keep track of new health and safety requirements. Thus changes to legislation etc. are initially researched by the corporate services group who pass on the relevant information to the different regions. The regional offices then check that each site complies, providing assistance as necessary.

As with some of the previous organisations, disaster recovery is now a major concern for the group, particularly in the light of recent fires at historic locations of national importance. These fires have persuaded the organisation that there is a real need to consider how they would deal with similar disasters at their sites. As a result, regional disaster officers have been appointed and a disaster procedures manual has been compiled for each major site. The latter includes prevention measures and the provision of salvage teams.

Strategic facilities management

As stated earlier, this organisation can really be regarded as a professional facilities management group. It comes perhaps as no shock, therefore, that the organisation actually places quite a lot of emphasis on strategic considerations. The organisation has recently produced a corporate planning document, which sets out the organisation's objectives for the next three years. The group's aims as laid out in the document can be basically summarised as below:

- to put and keep all of their properties into a condition appropriate to their importance, with regard to the urgency of the work and in accordance with the available resources;
- to prepare, review and update a defined basic minimum standard of documentation for all of their properties in order to assess their importance and condition;
- to make their properties accessible to the public, providing interpretation and facilities to make the visits enjoyable and informative in a way that reflects their relative importance as part of the national and international heritage;
- to play a leading role in the wider world, making use of their properties to demonstrate good practice and management, and to promote among others a commitment to conservation.

Now that the organisation has decided what it wants to achieve, the above aims have to be translated into action. Even a quick glance at the list suggests that this will not be an easy task, not only due to the number of factors involved but also because there are direct conflicts of interest. The organisation wishes to make their properties more accessible to the public through the provision of improved facilities, but an increase in visitor numbers could well have a detrimental effect on the building fabric. However, visitors are a major source of revenue and thus should be encouraged so that the organisation has more money to spend on preservation. The organisation, therefore, has to try to achieve a balance so that all of the aims can be realised as far as is feasible.

Overcoming these internal conflicts is a daunting task in itself, but the organisation understands that external forces will also have an influence on the way that they approach the various aims. For example, when they are trying to decide which properties have the greatest development potential, they have to consider what their competitors are doing, as well as deciding which sites are historically the most interesting or the easiest to build upon. They have to decide whether it is worth developing a site of great historical interest located miles away from any other attractions, when a less important property may benefit from passing trade to a newly improved competitor's site. The situation is further complicated by the fact that the organisation is not just competing for visitors with other heritage groups, but also with the rest of the ever-increasing leisure industry. Thus the organisation is being forced to take a much more commercial attitude than ever before to ensure that it attracts enough visitors to finance necessary maintenance and building work. To address this the organisation is planning to conduct a series of advertising campaigns in order to increase public awareness of its properties.

In addition, the historic properties group is already involved with other strategic initiatives in conjunction with the other groups within the larger organisation. For example, it has been agreed that the direct labour force currently employed in-house that carries out building and maintenance work is to be privatised within the next three years. It is anticipated that this move will make the labour force more efficient as it will actually have to compete for the work that it now gets automatically.

Comment

This organisation is responsible for the management of a considerable number of properties and hence in many ways it has quite well-developed systems, both operational and strategic, compared to the previous organisations. However, the organisation is not happy to stand still and is continually striving to improve its services, e.g. by actively seeking feedback from visitors.

This particular case highlights the fact that strategic facilities management is often a complicated balancing act. On the one side the facilities team have to consider what internal improvements are desired by the core business and users, whereas external forces may well be pushing the organisation in another direction. Thus a strategy has to be worked out that considers all of the different factors.

On a final point it is interesting to note the fourth aim listed above. This shows that certain organisations are beginning to realise that perhaps they have a duty to the world at large and that where possible they should share their knowledge so that others may follow their lead.

1.5.8 Public sector example 1

Case Study 7: NHS healthcare trust

This NHS trust situated in the North West of England provides a wide range of services to a multi-cultural and diverse population. The trust recognised that the cultural and economic diversification of its patients and visitors required services that are timely and sensitive in approach, ensuring the delivery of appropriate local and regional healthcare services. This organisation has been deemed as representing 'best practice' amongst the different trusts investigated during recently concluded research, on account of the mechanisms provided for managing performance measurement in facilities management.

Background

The trust, as a major service provider, was renowned as a national and, in some cases, international centre of excellence for healthcare and research. As a trust, it had a proven track record both in the delivery of quality care and in the development of students from all professions engaged in health. This was ultimately achieved through closely fostered relationships with the universities within the area. In partnership with the universities the trust had an excellent academic record and was committed to providing the highest standards of education, teaching, research and development.

The trust, as a major service provider, endeavoured to ensure that work by clinical and managerial staff continuously developed clinical services in order to remain at the forefront of healthcare delivery, research and teaching.

Transformation within the core operation demanded a similar transformation within facilities management, one that would reinforce the performance links and enable the organisation to gain competitive advantage. An internal service culture evolved and new facilities strategies emerged, more closely aligned with the objectives of the core organisation and more visibly connected with performance.

This trust provided exemplary insights into the generation of a performance measurement culture in facilities management, the alignment of facilities management functions to the core business and overall organisational effectiveness.

Facilities management structure

Estates and facilities are essential elements in the success of modernising the NHS. The NHS Plan emphasises the need for adequate capacity to treat and care for patients, by providing a modern high quality environment with modern systems of care. It also stresses the importance of the NHS becoming a better employer, including providing a modern working environment with good quality facilities for staff employed. Thus, facilities management issues are seen as vital to the success of the NHS Plan.

A robust estate strategy is essential to ensure that there are high quality well-located buildings, which are in the right condition to facilitate the delivery of modern patient care services. The benefits to a trust, and the wider health economy, of having a formal facilities and estates strategy include the provision of:

- an assurance that the quality of clinical services provided will be supported by a safe, secure and appropriate environment;
- a method of ensuring that capital investments reflect services strategies and plans;
- a plan for change that enables progress towards goals to be measured;
- a strategic context in which detailed business cases for all capital investment can be developed and evaluated, however funded;
- a clear statement by the trust to the public and staff that it has positive plans to maintain and improve services and facilities;
- a means by which a health authority can identify capital investment projects that will require its formal approval;
- a clear commitment to complying with sustainable development and environmental requirements and initiatives;
- an assurance that asset management costs are appropriate and that future investment is effectively targeted;
- assurance that risks are controlled and that investment is properly targeted to reduce risk; and
- a clear commitment that surplus assets are and will be identified over time and will be either disposed of or used for future service needs.

In this context, the trust's main strategic objectives could be listed as follows:

- *Obtaining best value* to agree, implement and deliver the facilities contribution to the relevant phase of the financial recovery plan.
- *Developing the estate* for the long-term investment to redevelop the site.
- *Developing the estate* using alternative funding mechanisms.
- *Minimising risk/controls assurance* to deliver upon corporate governance requirements.
- *Operational services* to continue to provide and develop high quality hotel and estate operational services in support of clinical care.

The trust was undertaking an extensive programme of reconstruction to its outdated buildings that will serve to enhance the clinical services located on the site through Healthcare in Partnership. The developments included the transfer of accident and emergency services from a neighbouring hospital, requiring the expansion of the accident and emergency department and the building of a new ward and day case theatre block to accommodate the transfer of services.

Staffing and structure of the facilities management organisation
Within the trust, the facilities directorate was made up from three departments and these were:

- PFI and Interim Strategy Project Management;
- Property and Estate Development; and
- Hotel and Estate Operational.

These departments were further divided into the following functions:

- domestic/linen/accommodation;
- portering/transport/receipt/despatch;
- medical electronics and maintenance;
- operational estates;
- printing services;
- security;
- catering services;
- car parking;
- patient services (hairdressing, chaplaincy);
- reprographic services; and
- receipt and distribution.

Private finance initiative and its involvement with facilities services
In the immediate future, the trust will be involved in the development of the new hospital. The new arrangement will aim to provide a range of traditional facilities management services but these will not include any clinical services. All facilities management services will be based on the output specification and will encourage generic workers who have the ability to undertake any reasonable tasks within the realms of facilities management. The generic team approach has been developed for the future to maximise operational efficiency through multi-skilling at the margin of the traditional service boundaries. This merging of cultures aims to ensure that facilities management staff become fully integrated members of ward and departmental teams.

The trust board had recognised that the success of these plans would depend completely on the full support of all their employees. To achieve this they have defined the following strategies:

- the establishment of clear lines of communication and liaison at all levels;
- a process of task rationalisation to re-engineer the traditional practices;
- a total review of the workforce skill base and the establishment of a planned approach for initial and ongoing retraining needs;
- the marketing of culture change with the emphasis on enhancing patient stay experience; and
- the remodelling of the measurement criteria and practices used to assess the level of quality and customer satisfaction with services.

Management of facilities management services
Some of the functions described above were retained in-house due to the constant demand for those services. Each of the in-house functional units was

responsible for carrying out work in its own area of expertise. Other facilities related services, such as cleaning and major building work, were outsourced due to their specialist nature or fluctuating demand. Regular checks were carried out to ensure that the work was consistent with the specified requirements.

Facilities management and external influences

In a climate of fast changes, the performance management and improvement agenda is a key part of modernisation of the Health Service. The need to measure and publish performance is twofold:

- to enable continual service improvement, through the provision of the best possible management information;
- to demonstrate accountability to the public and Parliament for the spending of public resources.

However, many trusts are finding it exceedingly difficult to respond to the multi-dimensional measurement systems. Excessive measurement has become an overhead for many trusts due to limited capability and resources to launch improvement programmes based on these measurements.

Strategic facilities management

It is important for facilities managers to have an influence on strategic decisions and to demonstrate the contribution that facilities make to the achievement of organisational objectives and business targets. To this effect, a clear message emerging from this case study was that management must acknowledge that facilities are a business resource. In order to respond to changing business practices, the range and scope of facility activities necessarily extends beyond merely providing technical solutions to problems arising to ensuring that facilities effectiveness is maximised and occupancy costs are minimised. Another important aspect of operational assets that demands they should be considered as a strategic resource is their impact on the financial performance of the organisation that owns or uses the assets.

This trust illustrated how, by focusing on the overall business objectives of an organisation, a facilities manager can manage its resources to complement the core organisation's long-term goals. It also highlighted the effectiveness of adopting a partnership approach. The trust further highlighted how the value of facilities management can be improved through strategy implementation and by satisfying related critical success factors representing the facilities management strategy within the organisation.

Comment

This case has provided evidence for the emergence of the trends in the facilities management organisation pertaining to the increase in performance measurement applications within facilities management and a focus on continuous development. The discussion also indicated that these trends have implications for the management of performance in the facilities management organisation.

Transformation within the core operation demanded a similar transformation within facilities management, one that would reinforce the performance links and enable the organisation to gain competitive advantage. An internal service culture evolved and new facilities strategies emerged, more closely aligned with the objectives of the core organisation and more visibly connected with performance. The evolution of facilities management within the trust needed to match the pace of change elsewhere within the organisation. The degree of success achieved by the facilities team in turning around the performance of the core organisation brought about a recognition of the potential for further gains to influence organisational performance in more ways than just financial outcomes.

1.5.9   *Public sector example 2*

Case Study 8: University

This case is a university based within the heart of the modern centre of a major city in the North East of England. This university was recognised as the most successful UK institution for widening participation in higher education and has received substantial government funding to enhance this work. There was a large estate and its efficient operation and maintenance was very challenging due to the scattered nature and varied age and suitability of premises, which was the major function of its facilities management department.

Background

This university claimed to have a wealth of creative ideas and initiatives for further business development and diversification, particularly around the concepts of the 'global university' and 'the virtual university'. The virtual university concept was embodied in a variety of developments, most notably in the 'University for Industry'. Various development schemes were placing the university at the centre of the virtual university movement and investment will continue to be made to consolidate and extend its position in these crucial fields, so becoming a university that facilitated life-long learning in the information age.

The university comprised different school areas: Arts Design and Media, Computing Engineering and Technology, Education, Sciences and Social Sciences and had been rated excellent in every single subject assessed by the Quality Assurance Agency in the past. Its strengths in research were developing rapidly, shown by a fourfold increase in research income over the last two years. Further, it was recognised as the most successful UK institution for widening the participation in higher education and has received substantial government funding to enhance this work.

Facilities management structure

The university's facilities management department was both manager of a substantial resource, the estate and facilities, the university's second most valuable asset, and was responsible for management and provision of a wide range of services essential to the development, operation, maintenance and care of premises. It was also a service, which by nature cares for students, staff and visitors of the university through a variety of personal contacts with the staff of its facilities department.

There was a large estate and its efficient operation and maintenance was very challenging due to the scattered nature and varied age and suitability of premises. The service was responsible for a significant proportion of the university's annual budget (approximately 10%) and for the management of substantial capital funding in relation to estate development and maintenance. The university's facilities management department was concerned with:

- taking care of students, staff and visitors of the university;
- creating a safe, secure and pleasant environment in which to work and live;
- ongoing review, updating and implementation of the university's accommodation strategy;
- general management of the estate;
- operation and maintenance of the estate; and
- provision of estate services.

Current structure

The facilities management structure of the university was reviewed and updated in 1996 in line with the university's preference for flatter structures and encouragement of devolved authority, responsibility, quality at point of delivery and greater customer focus. Its Facilities Management Division was structured into five service divisions. The service functions were separately identifiable for responsibility and accountability purposes, but the extent to which the Department was mutually dependent on individual and team contributions across these divisions cannot be overemphasised. The Facilities Management Department's divisions were as follows:

- projects;
- operations and maintenance;
- technical services;
- house services; and
- administration office.

Whilst concentrating on their own areas these divisions worked in very close cooperation and liaison with each other, particularly with respect to the commissioning of new buildings and other changes to the estate portfolio, which had service-wide implications. The divisions were very much interdependent in many respects and could only function efficiently with good interteam working. The divisions were supported by an administration office, which provided general administrative and clerical services support and undertook cross-service personnel and financial administration. External consultants were engaged to provide advice and services in specific subjects, in particular with regard to estate development planning, design of new buildings, major premises alterations and other specialist areas.

Management of facilities management services

The university's facilities management mission was 'to contribute to the aims and objectives by providing and caring for a quality environment in which to live, learn and work'. The service aimed to do this by:

- achieving customer satisfaction in all facilities and services provided;
- providing best value through applying innovation in the design, procurement and delivery of estate services;
- providing best professional advice in relation to the management and operation of the university estate;
- assisting, advising and providing data in relation to reviews of the accommodation strategy, space allocations and facilities provided in meeting the university's strategic priorities;
- being responsive to the operational needs of the university and the changing requirements of the university community;
- having regard at all times to the possible impact on the local community and on the environment of estate developments and the provision of facilities;
- complying with statutory requirements and relevant codes of good practice;
- promotion of a comprehensive quality management approach for the service and pursuit of a staff training and development programme to improve the capability and working performance standards of the service;
- ensuring alignment between institutional strategic priorities and the planning, design and delivery of estate services; and
- promoting a working culture that encouraged and developed individual and team contributions.

The facilities management plan of the university sought to facilitate and support the strategic directions of the university through an estate management strategy and a quality strategy.

Contracted maintenance continued to be tendered on an annual basis to ensure that best value for money and appropriate standards of service were maintained. The Facilities Management Department analysed all current maintenance contracts and rewrote specification documentation for re-tendering whenever required, and developed formal contract performance monitoring schemes to ensure that contracts let by the Facilities Management Department met the levels of service set out in the specifications and tender documentation, and hence provided value for money.

Meeting current core business needs

The university was promoting a comprehensive quality management approach for service and pursuit of a staff training and development programme underpinned by the 'Investors in People' initiative to improve individual and team working performance standards. In particular, the service had taken, and was taking, various actions to deliver appropriate staff training and development, refined policies, practices and procedures, improved client liaison and feedback, and pursued competitive pricing in the interests of efficiency, cost-effectiveness and achievement of client satisfaction.

Service-wide developments had resulted in the Department being acknowledged as delivering efficient services, providing value for money and being able to demonstrate this through a number of qualitative and quantitative evaluations. This included detailed examination of a number of sector-wide and external cost comparative benchmarks and performance indicators. Results of this exercise confirmed the Facilities Management Department to be performing well above average in all areas of service delivery.

Facilities management and external influences

The facilities of a Higher Education Institution (HEI) are one of its most valuable assets. They create a first impression of the organisation, so are key elements in marketing the institution (Cotts, 1990). A facilities strategy draws its aims from the institution's corporate plan and establishes the facilities needs to achieve these aims. HEIs operate in a global environment and compete for students and funds internationally, as well as in the UK, with the intention of maintaining or increasing their share of student numbers. In this competitive context, they face universal challenges to make efficient use of available resources and provide quality assurance.

Whilst the higher education properties can contribute to high quality education, it is the interrelationship within the organisational context that provides the catalyst for improved performance. From a business point of view, and from one of public accountability, the effective and efficient management and use of the property resource is imperative for all higher education institutions. Proliferation and diversity of technology and adaptation of sharing facilities (use of common teaching spaces etc.) and greater emphasis on quality in the study place are some of the potential implications of the changes for universities. Externally, they may inevitably suppress the demand for teaching spaces of universities. This in turn will increase the need to adapt redundant spaces to new uses. On the other hand, the recent massive expansion in higher education participation has forced universities to achieve more economic use of their facilities.

Special issues of facilities performance in higher educational organisations

Performance evaluation will play an ever increasing role in building design as external and internal factors place more demands upon the facility. Measuring performance explicitly focuses attention on feedback loops and this influences behaviour.

This is especially true for universities, such as that in this case study, which are entrusted with the responsibility of utilising public funds judiciously. Performance measures provide a mechanism to both learn from the past and evaluate contemporary trends in the use of facilities of universities. It is therefore hoped that the collection, interpretation and analysis of information about performance measures of facilities will provide the key to better planning and design for the future.

External comparisons

The service delivery was increasingly looking outside the sector for the exchange of comparable best practice and this strategy was to be a continuing feature throughout the service planning period. To this effect, the university reported in this case study continued to carry out annual benchmarking comparisons to inform the service on comparative service efficiency and value for money provided. This increasingly illustrated the high performance and low comparable cost of service provided.

There was a continuous trend of premises related revenue and expenditure for the university year on year, representing the continued increasing efficiency of the service in what was now clearly one of the most efficient in the sector. Recent capital underfunding of backlog major maintenance and redecoration programmes continued to be a matter of concern for the service. Current levels of maintenance expenditure needed to be improved otherwise a serious maintenance backlog would continue to expand over time.

Strategic facilities management

One of the most important developments within the university over the last decade has been the growing recognition of the strategic importance of facilities management. If facilities are perceived to be poor performers then this is not likely to enhance their chances of contributing to the strategic direction of the higher educational institution. Studies carried out by the university are increasingly reporting that management of the university is focusing attention on facilities improvement for a number of reasons, especially in a search for competitive advantage. Due to the nature of the activities, the background and the assignment, the university has a different view of these facilities management related activities and brings more of a sense of scientific enquiry to these activities.

Comment

The management and leadership challenge required to be met, at all levels of the university, was to reposition the service from its current *acceptable* level of performance and efficiency to a position of providing service *excellence*. This meant the provision of high quality services in the required areas, at the right time, closely matched to the needs of the clients and at the least possible cost. The delivery of facilities services to consistent standards was pursued, serving an increasingly demanding and diverse customer base. The scope to maintain previous years' cost-improvement savings had reached the stage where any new savings of a significant nature could only be delivered if matched against reduced service levels coinciding with reducing demands from service users or the cessation of delivery of some selected services.

Barriers to achieving change were not restricted to the areas of skill mix, expertise, experience, or staff training and development. Although these areas were very important, crucial factors were the values, attitudes, personal maturity and intellect of individuals and their ability to function as part of an effective team. These were the main characteristics that determined the operating culture within the Facilities Division of the university and were also at the core of such key quality issues as devolved management, individual responsibility and accountability, and ultimately effective and efficient team performance.

1.6 Conclusions

This chapter has tried to both present and illustrate a broad, dynamic, but coherent view of facilities management. It is hoped that this will help practitioners think boldly about how they can manage and enhance their operations. The next chapter homes in on a key issue for striving for excellence – a real client/user orientation.

References

Beer, S. (1985) *Diagnosing the System for Organizations*. John Wiley & Sons, Ltd, Chichester.

British Institute of Facilities Management (1999) *Survey of Facilities Managers' Responsibilities*. BIFM, Saffron Walden.

Cotts, D. (1990) Organising the department. Conference Papers of Facilities Management International, Glasgow.

Galbraith, J. (1973) *Designing Complex Organizations*. Addison-Wesley, Massachusetts.

Kast, F. and Rosenzweig, J. (1985) *Organization and Management: A Systems and Contingency Approach*. McGraw-Hill, New York.

Thomson, T. (1990) The essence of facilities management. *Facilities*, 8(8).

2 Excellence Through a Client/User Orientation

2.1 Introduction

2.1.1 Scope of the chapter

The aim of this chapter is to focus in on the issue of how the built environment impacts on users. The basis for this emphasis is that facilities management can only achieve excellence by better understanding and responding to the effects it has on the core business in the spaces managed, and this means on the occupants. This is put in the context of the huge scale of the built environment in the economy, but the low level of certainty about its value to users, despite many studies of individual aspects, such as the effects of air quality. In the face of this wealth of information, but lack of knowledge-for-practice, a way of looking at things from the viewpoint of the user is described. This builds on the simple premise that we experience our built environment spaces via our senses and balance out the information in our brains. The design implications of this perspective, and a case study of how this has elucidated school design, is provided.

2.1.2 Summary of the different sections

- Section 2.1. Introduction.
- Section 2.2. The place of the built environment in society and its scale in the economy is highlighted. The challenge is framed to look beyond base threshold levels and to develop holistic, integrated solutions from users' perspectives.
- Section 2.3. Discusses some general studies of the impacts of the built environment on users.
- Section 2.4. This section considers the evidence for the impacts of various aspects of the built environment.
- Section 2.5. Three design principles for user-orientated design are set out and illustrated with an empirical study of school designs.

Facilities Management: The Dynamics of Excellence, Third Edition. Peter Barrett and Edward Finch.
© 2014 John Wiley & Sons, Ltd. Published 2014 by John Wiley & Sons, Ltd.

2.2 Societal context

As indicated in Chapter 1, facilities management provides a key role in the built environment (BE) by working to maximise the value delivered by built assets to users and occupying organisations. To do this effectively those performing the facilities management function must be driven by a desire to really understand, and so be able to satisfy, the needs of the building users.

Looking from a building cycle perspective given in Figure 2.1 (Barrett, 2008), it can be seen that there may be good ideas in the design phase, but these only create 'potential' that has to be 'delivered', without too much erosion, through construction and then, crucially, 'realised' in the use phase. This is where facilities management has a strong, albeit rather reactive, role to play. However, FM can and should go further in two ways, related respectively to scope and phase. In terms of scope FM has a great potential to impact on the urban environment. This is less complex when an FM organisation has responsibility for an estate, but can be more involved when negotiations for mutual gain around shared responsibilities is required. The second element relates to the phase at which FM inputs and here there is significant opportunity to move from a reactive to a proactive relationship when buildings are being created or changed. Those performing the FM function should use their specialist knowledge and experience of working closely with occupants of buildings to bring this evidence centre-stage into the briefing and design process. By actively performing this role FM can be a catalyst for profound change in the orientation, and thus effectiveness, of the built

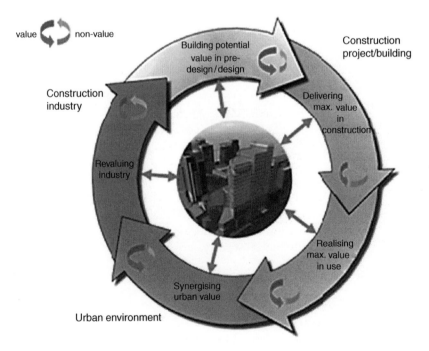

Figure 2.1   The building life-cycle.

environment professions. FM professionals can authentically and powerfully represent a drive for 'value for users'. This is the fundamental building block to an effective BE sector, which, for example, makes up around 20% of GDP in the UK (National Platform for the Built Environment, 2008). If this aspect is effectively addressed, then many other value opportunities can develop, such as improved rental values for offices where staff work more efficiently or improved delivery of healthcare in hospitals where patient recovery is hastened.

The argument for this approach may seem obvious, especially given that it has been estimated that people spend as much as 90% of their lives indoors (CIB Task Group 42, 2004). Thus, the potential impact of the quality of spaces on health and well-being is immense. Unfortunately, it appears that individual needs are often not being met for even the basic conditions of their living environments, such as temperature, lighting and air quality, with a number of consequential negative health effects (Bonnefoy *et al.*, 2004; World Health Organisation, 2007).

Currently, legislation and guidelines are used predominantly to define thresholds for discrete environmental factors: air quality (ventilation), thermal, acoustic and visual factors. Once basic standards have been met there is often no clear guide to priorities: indeed, evaluating the effects of the physical environment in relation to multiple factors is difficult owing to the interplay of factors (Higgins *et al.*, 2005: p. 36). Just because an indoor built environment is meeting current standards for environmental factors, it may not necessarily mean that a building is healthy, comfortable and safe from the users' and occupants' perspective. For example, most standards, such as for thermal comfort, are based on averaged data, thereby overlooking the fact that buildings, individuals and their activities may differ widely.

However, there is increasing awareness of the need to understand the particular needs of users (Saxon, 2006: p. 30) and the way they actually use their building, which may itself influence conditions (Bonnefoy *et al.*, 2004: p. 17). The complexity of these technical issues calls for an approach that aims to look beyond base threshold levels, to develop holistic, integrated solutions from users' perspectives.

In going beyond the issue of avoiding 'bad' spaces, there is the question of how much good space could contribute to society in terms of optimal human functioning. This logically lies at the heart of the assessment of the value of the built environment in society and should drive 'an awareness of the systemic contribution' that buildings (Barrett, 2008), and more specifically spaces, make to our lives.

This chapter seeks to open up these issues by illustrating some of the powerful effects the BE can have on users of buildings. Then the extensive, but fragmented, underlying evidence base of the sensory impacts of spaces will be reviewed. Lastly, a frame of reference will be proposed to focus this evidence into actionable design principles.

2.3 Built environment impacts on users

An extensive literature and practice-base has developed around 'building performance', with a wide variety of typologies on offer (Preiser and Vischer, 2005). A well-known example of this area of work is the PROBE series of case studies/user interviews that focused on building comfort issues (Leaman and Bordass, 1999). Post-occupancy performance (POE) studies of buildings have been widely advocated as a means for assessing how well buildings perform compared with the client/design aspiration. This has given rise to well-developed techniques for multi-method assessments to be carried out (Zeisel, 2006). The idea is that the intelligence will then feed forward into new designs. However, POEs are not commonplace and the lessons learnt are not generally available for use in practice (Bordass and Leaman, 2005). Another strand of development in recent years has been the rise in polemical works arguing for 'inside-out design' (Frank and Lepori, 2007) that builds from a focus on user needs and challenges the visual dominance of much design effort (Pallasmaa, 2009). This is twinned by those arguing specifically for aspects of sensory-sensitive design (Derval, 2010; Lehman, 2011).

These efforts stress that the evidence of building users' needs should be taken into account more fully. Copious case study examples illustrate the potential *elements* of 'good' design solutions. However, there remains a big gap between these putative elements and effectively achieving the desired *holistic* effects for users. Norman (1998) has highlighted the pervasive difficulty of designing 'everyday things' in a way that helps and supports users. However, some aspects have gained traction, for example Ulrich's (1984) classic evidence of the positive healing effects of views of nature. Less known, Mahnke (1996) provides comprehensive advice specifically on colour for various building types and it is interesting to see even now how designers find it hard to accept that 'white is not neutral' (Pernao, 2010). However, in any event these fascinating inputs still fall a long way short of comprehensively addressing the design challenge.

2.4   The complex sensory impacts of spaces

This section sets out to provide some illustrations of the complex dynamics of the various sense impressions people experience in spaces. Two themes are pursued throughout: firstly, the non-linear and relative nature of individuals' impressions and, secondly, the powerful impacts on human health and functioning. Finally, some complicating issues will be raised, including interactive effects between the various sense experiences, both physiologically and socially driven. Examples will be given concerning colour, lighting, layout, air quality and acoustics. In each case, general material is included, twinned with examples drawn from the relatively controlled environment of schools, where explicit measures of human performance, such as exam/test results exist.

2.4.1   Colour

Central to the impact of colour is the curvilinear issue of avoiding over- or understimulation through the degree of complexity or unity (uniformity) employed (Mahnke, 1996: pp. 22–27). There are many examples of the importance of difference and balance, rather than absolute colours, mediated to some extent by natural expectations, e.g. dark below and light above (Durao, 2000). In addition, laboratory experiments have shown that different colours can directly affect an individual's impression of, for example (Mahnke, 1996: pp. 72–75):

- temperature – light blue cooler; red/orange warmer;
- sounds – shrill, high pitched may be offset by olive green;
- size of objects – dark colours heavier; less saturated colours less dense;
- size of spaces – light or pale colours recede and so increase perception of size; dark or saturated hues protrude and decrease apparent size.

When selecting colours, the nature of the task is relevant. If concentration is sought, it can be assisted through the use of different colours, discriminating between those that are psychologically stimulating as opposed to physiologically stimulating. For example, in schools, cooler hues have been found to be good for concentration (Mahnke, 1996). Other studies have shown that intensity of colour, unity of colour schemes (Durao, 2000, p. 100) and contrasting end wall colour (Nuhfer, 1994) are also influential. Considering personal preferences, Heinrich (1980, 1992) carried out psychological colour tests on 10 000 children in age bands from 5 to 19 years and found, for each band, their preferred colours for their school environment. However, Heinrich points out that these colours are not always suitable for direct translation into wall colours.

2.4.2 *Lighting*

It is argued that natural daylight has a greater probability of maximizing visual performance compared with electric light because it tends to be delivered in large amounts, with a spectrum that ensures excellent colour rendering (Boyce *et al.*, 2003: p. 65). Evidence from research on the impact of design in hospitals shows outcomes related to the variables of nature and light. Studies in a Swedish hospital on heart surgery patients showed that the viewing of nature reduced patients' stress and need for analgesia compared with those in control groups (Ulrich, 1984, 1991; Ulrich *et al.*, 2004). However, Boyce *et al.* (2003) argue that the 'biophilia hypothesis' (we like nature) ought to be investigated in order to understand better the role of windows and the relative impact of a view out versus letting daylight in (p. 68). Whatever the scenario, they maintain that the effectiveness of daylighting (and artificial lighting) will depend on how it is delivered (p. 65), linking to issues of glare and distraction. From an occupant's perspective, there is often a battle between the desire to be close to a window versus the problem of glare (Christoffersen *et al.*, 2000).

When daylighting is not available, preferred lighting levels have been shown to be higher than indoor lighting standards and may correspond to levels where biological stimulation can occur (Begemann *et al.*, 1996). Special artificial lighting that emulates the full spectrum of sunlight can be used to tap into natural circadian rhythms with powerful impacts on involuntary emotions (Rea, Figueiro and Bullough, 2002; Joseph, 2006). For example, Figueiro *et al.* (2003) have drawn on the physical fact that blue light has the maximal effect on circadian rhythms to work with Alzheimer's patients to improve their sleeping patterns through exposure to quite low levels (300 lux) of blue light for two hours before the patients' normal time for going to sleep. It was found that this level of exposure shifted the patients' level of activity towards the day and away from the night, delaying the decline of body temperatures by two hours and leading to enhanced sleeping between 2.00 am and 4.00 am.

Looking more specifically at schools, the Heschong Mahone Group (1999) studied the impact of daylighting on learning in schools. They looked at 21 000 elementary school pupils and classified 2000 classrooms for their daylighting levels. Controlling all other influences and using multiple linear regression analysis, they found positive correlations between the variables: that students with the most daylight progressed 20% faster in mathematics and 26% faster in reading compared than those with the least. However, this finding was reversed when there was glare from certain designs of skylight. Furthermore, another study by the same group in 2003 (Heschong Mahone Group, 2003), of different schools in a different climate, found daylight was not significant in predicting performance. Further analysis showed that other characteristics associated with daylighting, which did not exist in the earlier studies, such as noise, were affecting performance through a counteracting negative effect. This highlights both the strength of the potential impacts and their fragility, especially in combination with other factors.

2.4.3 *Layout*

Issues of layout have been shown to affect behaviour; for example, aggression and destructive behaviour increase as the size of the classroom and number of children in it increases (Rivlin and Wolfe, 1972). High-density situations (too many children or inadequate space) lead to excessive levels of stimulation, stress and arousal, reductions in desired privacy levels and loss of control (Wohlwill and Van Vliet, 1985). Academic performance suffers (Achilles, Finn and Bain, 1998) as students' sense of responsibility and meaningful participation is discouraged (Moore and Lackney, 1994). Within this context, (Marx , Fuhrer and Hartig, 1999) found that semi-circular seating arrangements engendered a significantly higher, and enduring, level of question-asking in fourth graders compared with row-and-column arrangements.

Clearly marked pathways through the building and within classrooms improve the utilization of spaces, but more importantly they help keep the children orientated and stimulate their imaginations (Alexander, 1977). More tangibly, school buildings and grounds with 'clearly defined areas for freedom and movement' have been found to correlate significantly with higher student ITBS (Iowa test for basic skills) scores (Tanner, 2000: p. 326). Indeed, student activities in naturally planted 'school yards' have been shown to be more creative than in classrooms or traditional playgrounds (Lindholm, 1995), with positive effects on learning and cognitive qualities (Fjortoft and Sageie, 1999; Fjortoft, 2004). As an example of interaction between navigation and colour, it has been found that the use of colour to assist way-finding is especially important in primary schools (Engelbrecht, 2003).

2.4.4 Air quality

Perceptions of air quality are not absolute. There are varying decay curves for the perception of different smells depending on their source. For example, in laboratory conditions, adaptation to natural human body smells is quicker than to tobacco or building smells (Gunnarsen and Santos, 1999). However, perceived air quality cannot be the only measure for health effects as some toxic pollutants, such as carbon monoxide and radon, are not perceived (Berglund *et al.*, 1999).

A study of a primary school class by Coley and Greeves (2004) investigated the effects of low ventilation rates on their cognitive functions. The research used a range of computerised tests and showed that the attentional processes of the children were significantly slower, by approximately 5%, when the level of carbon dioxide (CO_2) in the classroom was high. Other studies have found a statistically significant association with CO_2 levels and absence levels from school. However, this is thought to be an indicator of low ventilation rates leading to increases in communicable respiratory illnesses (Fisk, 2000).

2.4.5 Acoustics

The multi-dimensional acoustical experience in, for example, performance spaces exhibits two interesting tensions. One is experienced by the audience, between clarity and reverberance, which tend to act against each other (Hidaka and Nishihara, 2004). The second tension is between desirable listening conditions for performers and audience. Performers need to feel 'support' from the hall, and so would like the most information-rich sound reflections returned quickly to the stage, while the audience wants all this important energy directed towards them (Jeon and Barron, 2005).

Evans and Lepore (1993) have isolated the negative effect of noise on children's recall. They studied 1358 children aged 12–14 years in their own classrooms, using standard tests, but under different noise conditions. They then tested the pupils' recall a week later and found that a statistically significant decline in performance could be associated with the noisy conditions. This negative effect has also been shown to affect long-term recall in children (Hygge, 2003).

2.4.6 Complexity and interaction

The individual's sense experiences clearly tend to be relative and changing. However, it can be seen that they are also interactive. Santos and Gunnarsen (1999) have specifically studied this phenomenon for an individual's choice

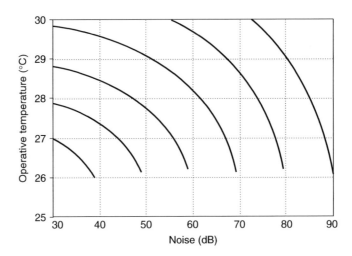

Figure 2.2   Iso-annoyance curves. Reprinted with permission.

between the noise of air conditioning (in a hot country) against being too hot. By working with subjects they mapped 'iso-annoyance curves' that represent where people trade between these alternatives (see Figure 2.2).

The human environmental sensory experience is further complicated by changes in a person's sense organs over time. For example, the ageing of eye lenses requires higher light levels to function well (Joseph, 2006), which makes colour discrimination more difficult (Durao, 2000: p. 113). Furthermore, the problem of glare intensifies (European Commission, 2003). In addition, the capacity and qualities of the brain varies over a lifetime (Carter, 1998: p. 21), with changes through potentiation or pruning of connections occurring through learning (Goswami, 2004). For example, the capacity of a child's brain shows less flexibility than an adult's in processing changes in spatial categories (Hund and Plumert, 2004), but the elderly (Sundstrom *et al.*, 1996) and those with Alzheimer's disease also have problems with spatial processing and way-finding (Zeisel, 2006).

The way we experience our environments through our senses is complex and the impacts on our behaviour, health and performance are profound. These responses are particularly hard to predict in the face of interacting influences. Here, humans must make balancing calculations in their brains, which is the focus of the next section.

2.5   Focusing through user-orientated design principles

Linking the above multiple aspects into coherent design for supportive facilities is a considerable challenge. Nasar (1999) provides an extensive synthesis of the factors to consider in architectural design competitions, but in the process he highlights the disjuncture between architects' design preferences and those of 'normal' building users. This raises the question of the choice of the overarching perspective to synthesise the available alternatives into an optimal design.

One emergent way forward is to use the notion that as the user's brain is the place that resolves the multiple sensory inputs for that individual, these mental mechanisms can provide a basis for understanding the combined effects of sensory inputs on users of buildings. This is the approach being championed by the Academy of Neuroscience for Architecture (ANFA), based in San Diego, and underpinned by works such as Eberhard's book (Eberhard, 2007).

Up to now the only exemplar study using this sort of thinking has focused on Alzheimer's care facilities (Zeisel *et al.*, 2003). Building out from an understanding of the impaired functionality of the patients' brains, this study has successfully shown how characteristics of the built environment can have medically convincing (but non-pharmacological) impacts on patients' symptoms. The study looked at the impacts on the well-being of patients in a range of holistically designed care environments in the United States and found clear clinical evidence of improvements in aggression, depression and withdrawal, when compared with other facilities. The sort of design features Zeisel *et al.* considered are pretty practical: e.g. providing individual privacy with treasured objects, maintaining a residential scale in design, providing common areas with clear behavioural cues and support for autonomy. The last of these relates to nursing care programmes tailored to each individual – facilities management of a high order! Although the factors are quite practical they are comprehensive and driven by a holistic view of the patients' needs (see Table 2.1). Figure 2.3 provides a typical view of one of the care facilities designed following this approach.

2.5.1 *Mediating neuroscience processes*

To understand the relationship between human sense experiences and their powerful effects on human functioning better, we can turn to evidence from neuroscience for clues as to the mediation performed through brain functioning. The complexity evident is rooted in the fact that individuals, in practice, experience spaces holistically and interactively. At a base level this is confounded by the cognitive limits of humans, so that perception becomes an 'ill-posed question', in which the brain endeavours to represent reality probabilistically, as best it can using Gestalt grouping rules (Wolfe *et al.*, 2006) correcting the stimuli into usable experience, called 'percepts' (Eberhard, 2007). These perceptions are intricately linked to memory, which relies on a type of matching and recognition of sensory information circuits (p. 106). Eberhard (2007) describes how our individual reactions to particular spaces, such as a sunny room, draw from these often unconscious memories (p. 45). He argues that when we experience a new sensation, such as when a child visits a hospital for the first time, all the senses will be at work trying to associate the new sensation with the memory cluster; the designer can then work to provide spaces that can be associated with a more pleasant experience (p. 71).

Table 2.1   Designing for Alzheimer's patients linked to three themes.

Feature	Description	Theme(s)
Exit control doors	These are designed to blend into the setting and have coded pushpads to limit residents' access to unsafe areas outside the residence without upsetting them and creating catastrophic reactions	Stimulation
Walking paths	Visually engaging paths with easily identifiable destinations transform aimless wandering and pacing into purposeful walkingm thus encouraging residents to move around the residence safely and independently	Stimulation Individualization
Individual privacy	Providing residents with both privacy and the opportunity to be surrounded by familiar furniture, décor, and treasured objects aids them in maintaining a sense of self and improves their recollection and memory	Individualization
Common spaces	Common areas designed to accommodate residential elements and spatial difference – with special attention given to kitchen, living room and activity spaces – provide residents with the cues they need to behave appropriately	Stimulation
Outdoor freedom	Residents' access to safe and supportive healing gardens provides them with cues to time of day and seasons which reduces disorientation and sun-dowing as well as comprehensible stimulation and a healing relationship to light, fresh outdoor air and the natural world	Naturalness Stimulation
Residential scale	All rooms and furnishings as well as programmes are designed to reflect a residential setting – to feel 'homey' – reinforcing hard-wired brain elements that reduce anxiety and aggression	Stimulation
Autonomy support	Both environment and programmes encourage residents to use their remaining abilities to carry out daily routines and support them in doing so	Individualization
Sensory stimulation	Sounds, sights and smells are managed so sensory stimulation is familiar and meaningful, over-stimulation is avoided, and total reliance on often-damaged verbal skills is avoided	Stimulation Naturalness

Specifically, the issue of layout and the navigation between and within spaces provides a good example of the complex, multi-dimensional way in which data are used and the resulting consequences. Neuroscience studies reveal that the brain operates using three different, but ideally complementary, systems for way-finding. These are:

- reading and interpreting signs;
- remembering routes as part of habitual behaviour, which can be part of our familiarity and attachment with a place (Quayle and van der Lieck, 1997);
- tracking landmarks in relation to the body (egocentric) or relative to other landmarks (allocentric), using the brain's own mapping system (Hartley, Trinkler and Burgess, 2003).

Figure 2.3   Hearthstone Alzheimer Care Residence, Woburn, Massachusetts.

For any particular space, individuals bring different background experiences and the design of particular spaces may not send coherent messages across the three mechanisms. The result in everyday experience is that way-finding in some spaces is intuitive, while in others confusion reigns.

Underpinning these processes are attentional mechanisms that give priority to one sense over another, and also help in the construction of the coherent image (Wolfe *et al.*, 2006: p. 179; Rolls, 2007: p. 436). The various human senses pick up information from our environment and this is processed in the brain in a variety of ways that lead to behaviour based on a mixture of responses, which can be broadly categorized as reflexes/autonomic, implicit or explicit actions (Rolls, 2007: p. 413). As will be seen below, the role of implicit/unconscious responses is particularly interesting in the context of the multi-sensory experience of spaces, as they subtly influence our explicit thought processes, but also because these effects are themselves powerful, albeit often overlooked.

The sequential process from sense input to response clarifies how all these competing influences interact and how the human brain seeks to make optimal judgements between alternatives. Rolls argues that human behaviour is ultimately motivated by 'primary reinforcers', drawn from our external experience, which are related to survival needs, such as: for clean unputrified air, bounded temperature, the absence of natural dangers, light, shelter, reasonable stimulation and food hoards (Rolls, 2007: pp. 18–19). This sensory information about the world is collected as raw data that then enters the orbitofrontal cortex of our brain where the value of the environmental stimulus is assessed. This appears to happen by a pattern-matching process (p. 148) against alternative strings of neuronal associations that are built up and progressively updated. This individual learning process links the elements of situations observed to the built-in primary reinforcers, so giving previously neutral inputs reward value as 'secondary reinforcers' (pp. 62–67), e.g. the sight of food rather than its taste. This reward calculation then feeds forward to the amygdala where individuals hold a sense of the correlation between the prospective actions and the potential rewards.

From here, the information feeds forward to the basal ganglia where inputs from various parts of the brain calling for action are weighed. These result in signals to motor regions and thus behaviour. The explicit system, rooted in the advanced language capacity of the human brain, enables implicit actions to be deferred for reasons connected to longer-term plans (Rolls, 2007: p. 422). Thus, when experiencing spaces, we will receive a range of sensory inputs that will create an implicit or unconscious response that may or may not be in line with our explicit reaction.

It seems that the implicit response alone does not automatically create behaviour in humans. Although some primary goals may be specified in our genetic material to optimize our behaviour for survival and reproduction, the actions to achieve these goals are not specified, thus allowing for lifetime learning and flexibility (p. 426). However, the implicit responses, represented in a range of relatively few emotional states or moods, do have an important impact by providing quite weak 'back projections', so influencing the cognitive evaluation of what is experienced or remembered. Thus, in situations where rapid responses are needed or there are too many factors to account for using the explicit system, then the balance may shift more to the implicit system to guide our response (pp. 198, 415). Furthermore, it would seem that moods are related in our memories to where events happened and so the connection between experiences and spaces is an important element of our mood and its impact on our explicit functioning (pp. 196–197).

2.5.2   Design implications

The above discussion gives some indication of the range and depth of the research evidence on human sense perception, impacts and the underlying cognitive calculations involved. Building on this and to support application in practice, it is proposed that three design themes can be seen to emerge:

- the role of naturalness;
- the opportunity for individualisation;
- appropriate levels of stimulation.

The rationale for the choice of these themes is summarised as follows. Firstly, as our emotional systems have evolved over the millennia in response to our natural environment, it does not seem unreasonable to suggest that our comfort is likely to be rooted in key dimensions of 'naturalness' that should, therefore, infuse the design process. The stress here is, of course, on the positive aspects of naturalness, such as clean air. At the other extreme, it is known that supernormal stimuli, such as the noise from man-made artefacts, for example cars and guns, can produce superstrong emotions because the stimuli are much more intense (unnatural) than those in which our present emotional systems evolved and that the subsequent responses are not necessarily adaptive (Rolls, 2007: p. 450).

For schools, significant evidence of positive impacts deriving from aspects of naturalness have been given above, especially around daylighting and planting. For healthcare, planting has been shown to aid recovery of the capacity to focus attention (Kaplan, 1995) and can be contrasted with performance in attentional tests in urban settings (Hartig *et al.*, 2003).

Secondly, the brain functioning described in the section above highlights the personal way in which individuals build connections between primary reinforcers and complex representations of secondary reinforcers. Taken together with the situated nature of memory, these personal-value profiles lead to highly individual responses to space. This provides a sound basis to raise the potential importance of 'individualisation' as an additional, key, underlying design principle. This appears to play out in two ways: particularisation and personalisation. Particularisation concerns accommodating the functional needs of very specific types of users, e.g. learning and way-finding in the context of age and physical requirements (Zeisel, 2006). Not surprisingly, responding to the needs of up to four genera-tions in the workplace has been highlighted as a complex and growing issue (Haynes, 2011). Personalisation concerns an individual's preferences resulting from their personal life experiences of spaces. These, of course, will vary greatly from person to person, but the desire is evident in the way people seek to individualise spaces.

Thirdly, lying behind the detail of design elements for general and particular needs, there is also a recurrent theme around the general level of stimulation that is appropriate for given situations. In broad terms this may vary from buildings designed for relaxation, such as homes, to those designed to stimulate, such as theatres, but also variation will be appropriate within buildings. Therefore, in a school, classrooms may need a different approach from assem-bly areas. More specifically still, focusing on colour, Mahnke (1996) proposes colour palettes for school classrooms that vary every three years to gradually reduce the level of stimulation and increasingly engender concentration as the pupils get older (pp. 183–184). Thus, a link can be seen here with the issue of individualisation, stressing the holistic nature of design solutions. In the context of judging design competitions, Nasar (1999: pp. 77–85) reinforces the central importance of the level of stimulation produced. Drawing from an extensive literature review, he suggests that combinations of pleasantness (or unpleasantness) and different levels of arousal yield either excitement (or boredom) or relaxation (or distress).

These three user-driven themes are shown in Figure 2.4 and linked to three aspects of the environment that demand attention. Having set out these themes that, it is argued, should inform design, it is encouraging to see how these have been successfully operationalised in environments that demonstrably support Alzheimer's patients. Taking the successful designs from Zeisal *et al.*'s study (mentioned above), these exhibited the set of practical and coherent design features summarised in Table 2.1.

It can be seen that the themes of (positive) naturalness, individualisation and appropriate (low in some places – higher in others) levels of stimulation pervade this design toolkit. It is also evident that complex practical issues are addressed alongside compounding psychological and sociological issues associated with spaces, such as status, control and social belonging (Vischer, 2005). Although,

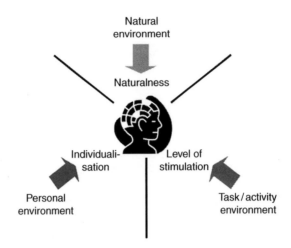

Figure 2.4   Three neuroscience-derived design principles.

or perhaps because, this example is centred on a particularly vulnerable group, it can be argued that the needs addressed represent a raw exposition of universal human needs.

2.5.3   HEAD case study testing propositions on holistic sensory design

The above example gives some idea that the three design principles (naturalness, individualisation and level of stimulation) may help guide us towards designs that respond to users' needs. However, further proof that this approach helps would of course be welcome, especially if it related to human performance dimensions that are seen to matter in practice.

Therefore, a research study was designed that, at a general level, took a multi-dimensional, holistic view of the built environment within which humans live and work and sought to discover its impacts on human well-being and performance. This is a complex and current issue, with no consensus currently as to the relative importance of internal environmental quality (IEQ) factors for overall satisfaction.

It was in this context that a research project was designed to address the impact of school design on the learning rates of pupils (see Endnote). In fact the aim of the HEAD (Holistic Evidence and Design: sensory impacts, practical outcomes) project was: to explore whether there is any evidence for demonstrable impacts of *school* building design on the learning rates of pupils in primary schools. Primary schools were selected as the pupils tend to remain in the same classroom for much of the day and there are available accepted measures of their academic performance (SATs) that are of great interest to teachers and parents alike.

The shape of the research endeavour is shown in Figure 2.5. To take this forward, hypotheses as to positive impacts on learning were developed for

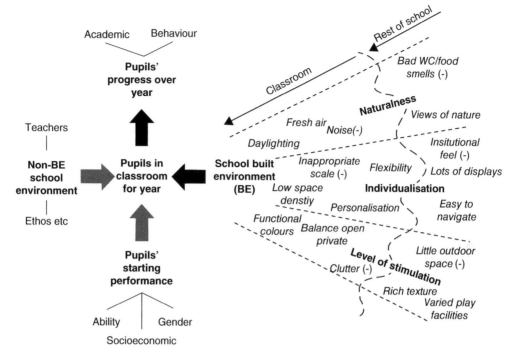

Figure 2.5 Schematic contextualising the impact of the school built environment on pupils.

ten design parameters (light, colour, etc.) within the neuroscience-derived framework of three design principles (described above). The hypotheses about the ideal design parameters are given in Table 2.2.

Data on classrooms and about the pupils in those spaces were collected, with significant assistance from Blackpool Council. This allowed the hypotheses to be tested using data collected on 751 pupils from 34 varied classrooms in seven different schools. The academic results are available in a journal paper (Barrett *et al.*, 2013), but in summary, the analysis employed a multi-level statistical modelling approach as this could reflect the 'nested' structure of the data (pupil in class). This enabled the unmeasured 'pupil effects' and 'class effects' to be partitioned as residuals at each level, so avoiding misleading results owing to the overestimation of significance that a simple regression analysis would deliver.

Overall, the model explains 51% of the variability in the learning improvements of the pupils over the course of a year. However, its explanatory power is asymmetrical across the levels of 'pupil' and 'class'. The multi-level approach identified a 73% reduction in the random error at the 'class' level, linked entirely to the design parameters in the model. Thus, only a relatively small random error remains at this level. As the 'class' level of analysis was the focus of this study, and is key to facilities management, the high level of explanation attaching to the design parameters is of great interest.

Table 2.2 E-H-P factors model.

Design principles	Design parameters		Indicators		Factors	Classroom characteristics making up high ratings
Naturalness	N1	Light	A	The quality and quantity of natural light the classroom can receive	1	Orientation of the room facing	Daylight can penetrate into the room from more than one orientation and the south side is towards the sun's path for most of the year
					2	Glazing area / floor area	The classroom can receive more daylight if the ratio is higher
					3	The most distant point from the glazing	The distribution of daylight level can be more even when this value is smaller
			B	The degree to which the lighting level can be controlled manually	4	Quality of the electrical lighting	More electrical lighting with higher quality can provide better visual environment
					5	Shading covering control	The blinds (shading coverings) are better than the curtains; All blinds (shading coverings) are in good condition; The space adjacent to the window is clear
	N2	Sound	C	The frequency of the noise source's disturbance	6	Noise from the school outside	The room is far away from the road traffic and there is a buffer zone between the room and traffic road
					7	Noise from the school inside	The windows are towards the quite area; there is no busy activity area adjacent to the room; The chairs have rubber feet
			D	The degree to which the pupils can hear clearly what the teachers say	8	Size and shape (length/width)	It is easier for teachers to locate pupils' seating position when concentration is needed in a rectangular classroom than a square one
					9	Carpet area of the room	More carpet area is, less reverberation time (RT) can be
	N3	Temperature	E	The degree to which the pupil feel comfort in summer and winter	10	Amount of the sun heat	Rooms with south façade can receive more sun heat than any other orientated rooms
			F	The quality of the central heating system	11	Heating control	Underfloor heating is better when it comes to evenly distribute the heat with a thermostat
	N4	Air quality	G	The frequency of the contaminated air that comes into the classroom	12	Contaminated air inside the classroom	Usually, CO_2 level is lower if the room volume is bigger when same amount of people in it
					13	Contaminated air from other spaces	The room is far away from the polluted air, e.g. toilet

			H	The degree to which the stuffy feeling can be adjusted manually	14	Opening size	The air exchange is quicker when the opening size is bigger
					15	Opening options	Different opening positions can give occupants more choices to increase the air movement
Individualisation	I1	Choice	I	The degree to which the distinct characteristics of the classroom allow the sense of ownership	16	'This is our classroom!'	Any design features that distinct characteristics of the room allow the sense of ownership
			J	The degree to which the FFE are comfortable and familiar, supporting the learning and teaching	17	Furniture, Fixture and Equipment (FF&E) quality	The facilities are comfortable with high quality, supporting the learning activities.
					18	Quality of the chairs and desks	The desks and chairs are comfortable, interesting and ergonomic
	I2	Flexibility	K	The degree to which the pupils live together without crowding each other	19	Size for the pupil's activity area	Bigger size helps pupil to learn better
			L	The degree to which the room plan allows varied learning methods and activities	20	Configuration changed to fit the size of class	Easier the teacher change the space configuration, more teaching methods can be adapted to pupils learning.
					21	Zones for varied learning activities	More zones can allow varied learning activities at the same time.
					22	Attractive (or useful) space attached to the classroom	The storage and/or breakout space are always available and not used for other purposes.
	I3	Connection	M	The presence of wide and clear pathway and orienting objects with identifiable destinations	23	Corridor usage	It is not used for storage and/or breakout purpose.
					24	Corridor width	Wider the corridor is, quicker the movement can be.
			N	Clear and orienting corridor	25	Clear and orienting corridor	Large and visible pictures and or landmarks are along the pathway.
					26	Safe and quick access to the school facility	The room is near the main entrance and other specialist rooms, e.g. library, music, café etc.
Stimulationn appropriate level of	S1	Complexity	O	The degree to which the school provide appropriate diversity (novelty)	27	Site area / total pupils in school	Bigger the site area is, more potential opportunities for the school to provide varied outdoor learning patterns and activities.
					28	Building area / total pupils in school	Bigger the building area is, more potential opportunities for the school to provide alternative learning rooms and spaces.

(Continued)

Table 2.2 (Cont'd).

Design principles	Design parameters		Indicators		Factors	Classroom characteristics making up high ratings
		P	The degree to which the classroom provide appropriate diversity (novelty)	29	Diversity (novelty)	The interior decor can catch the pupils' attention and arousal, but in balance with a degree of order. Diversity and/or atypicality are expected to be good in producing stimulation.
				30	Quality of the display	The displays are stimulating, well designed and organized, ideally without cluttered noisy feelings. Diversity and/or atypicality are expected to be good in producing stimulation.
	S2 Colour	Q	The degree to which the 'colour mood' appropriate for the learning and teaching	31	Colour of the classroom	Carefully considered colours for the wall and floor area. Taking age into consideration, warm colours may complement the young pupil's extroverted nature, while cool colours enhance the ability to concentrate on learning later.
				32	Colour of the furniture	Carefully considered colours for the furniture. Pupil age is also taken into consideration (same as 32).
				33	Colour of the display	Carefully considered colours for the display. Pupil age is also taken into consideration (same as 32).
	S3 Texture	R	The degree to which the views of nature through the window	34	Distant view	There is a wide-field vision with sky, distant urban and rural area and landscape.
				35	Close view	There are full of natural elements, e.g. grass, garden, pond, tree etc.
		S	The presence of fabric variety, seasonal cycles and varied learning opportunities	36	Outdoor play quality	The pupils can have abundant play area outside, ideally adjacent to the classroom.
				37	Outdoor learning alternative	The pupils can have varied learning opportunities other than in the classroom.

Six of the original ten built environment 'design parameters' were identified as being particularly influential in the multi-level model. The six parameters are: colour (18%), choice (10%), connection (26%), complexity (17%), flexibility (17%) and light (12%), with their proportionate influence (summing to 100%) indicated in brackets. It is interesting to note that there is a relatively even spread of influence across all six factors. This resonates with the notion that the impact of an environment on a user is a composite response via all their senses. There is a fairly even mix of aspects that are either design related or mainly use related, indicating that both designers and users have significant opportunities to take these findings into account in their choices about classroom spaces. This leaves a significant 'design' challenge to resolve competing requirements, but this study gives examples of 'good' and 'bad' spaces for each of the six parameters.

By fixing all the variables at their average values, except for the environmental factors, the model was used to predict the weighted progress (pupil's learning progression), owing to the environmental factors only. This in effect took an average pupil with an average teacher and placed them in each of the 32 classrooms studied. Comparing the 'worst' and 'best' classrooms, the environmental factors alone were found to have an impact of 11 points in learning progression, summed across three subjects. This suggests that placing the same pupil in the 'best' rather than the 'worst' classroom would have an impact on their learning that equates to the typical progress of a pupil over one year (11 points). Using the range in pupils' improvement in this data set, it was also possible to estimate the proportionate impact of the environmental factors on learning progression, in the context of all influences together. This scaled at a 25% contribution on average.

It should be remembered that we are looking at the spaces in functional terms, focusing entirely on the impact of the differences between spaces on the academic performance of the pupils. In this context it can be seen that parameters to do with the design principle of *individualisation* are prominent. Here the issue of connection (way-finding) has raised some surprising issues compared with prevalent theory, but these can be seen to make sense if a pupil's perspective is taken.

Addressing the principle of *appropriate level of stimulation* for learning is also important and raises the issue of functional requirements versus aesthetic preferences. Young pupils may like exciting spaces, but to learn it would seem they need relatively ordered spaces, with a reasonable degree of interest.

In the area of the principle of *naturalness*, only the parameter of light remained in the multi-level model, and even this was quite a complex relationship between a desire for light, a dislike of glare and the importance of good artificial lighting. Indeed, it was very commonly observed that blinds were closed to facilitate the use of whiteboards and projectors. All of the other environmental factors were found to be individually significant, but are not in the model mainly because, with this data set, they are quite extensively correlated with other design parameters, albeit at a low level. The effect of this is that these factors were competed out of the regression analysis. At a practical level it could be that air quality is less evident because it was found to be almost universally *poor* based on CO_2 spot checks in the classrooms. At the other extreme, it could be that temperature and sound (as one vote

Table 2.3   Features of classroom that best supports learning.

Design principle	Design parameter		Good classroom features
Naturalness	Light	*	Classroom receives natural light from more than one orientation and (or) natural light can penetrate into the south windows
		*	Classroom has high quality and quantity of the electrical lightings
		*	The space adjacent to the window is clear without obstruction
Individualisation	Choice	*	Classroom has high quality and purpose-designed furniture fixture and equipment (FF&E)
		*	Interesting (shape and colour) and ergonomic tables and chairs
	Flexibility	**	More zones can allow varied learning activities at the same time
		**	The teacher can easily change the space configuration
	Connection	*	Wide corridor can ease the movement
		***	The pathway has clear way-finding characteristics
Stimulation,	Complexity	*	Big building area can provide diverse opportunities for alternative learning activities
appropriate level of		*	With regard to the display and decoration, classroom needs to be designed with a quiet visual environment, balanced with a certain level of complexity
	Colour	*	Warm colour is welcomed in senior grade's classrooms while cool colour in junior grades, as long as it is bright
		**	Colour of the wall, carpet, furniture and display can all contribute to the colour scheme of a classroom. However, it is the room colour (wall and floor) that plays the most important role

*Design related classroom features.
**Usage related classroom features.
***Future study is needed to pursue its positive characteristics.

veto factors) are important but did not rise to the top of the analysis as they simply *have* to be addressed by users and so are not allowed to get very poor in practice.

In summary, a significant impact has been revealed on learning that has been isolated to the built environment provided. Classrooms designed in particular ways that address naturalness, individualisation and the appropriate level of stimulation do lead to enhanced progress in pupils' learning. A summary of the main design parameters critical to a good learning environment is given in Table 2.3.

It can be seen that there is a mix of factors that can be roughly defined as design related and/or use related. This raises interesting opportunities in facilities management terms to affect designs, but also to optimise the configurations of existing spaces. This may sound quite theoretical, but in fact is all linked to practical spaces studied. This is made clear by focusing on the classrooms that provided the least and most effective learning environments, selected from the 34 classrooms studied (solely in terms of impacts on learning rates). Of course the worst does not include the poorest aspects on all fronts, nor the best all the most exemplary.

The salient features of the *least* effective learning environment were:

1. Windows on one side, with the blinds always closed or half closed.
2. Bookshelf fixed along the window wall, which is a disincentive to manual operation of the blinds and window.
3. Poor FF&E with no sink or cupboard, or smart control station.
4. No dedicated PC corner.
5. Small classroom size and few learning zones with little flexibility.
6. Within high density building where the pupils do not have many alternative spaces for learning.
7. Noisy visual environment.

There are other theoretically *poor* aspects of this classroom, although these were not prioritised by the statistical analysis, namely:

8. No view outside due to high window height.
9. No distant view either as it is adjacent to another building.
10. Small outdoor space, located in the urban area.

Now turning to the *most* effective learning environment from amongst the sample of classrooms studied (see Figure 2.6), the salient features of this *excellent* learning environment are:

1. Large windows with glazing towards two orientations (the blinds happen to be down in the figure!).
2. No obstructions along the windows. Both teachers and pupils can easily use the windows and shading covering (one side).
3. High quality and quantity of the flicker-free electrical light.
4. Good FF&E, e.g. big smart board, centralized teacher station, PC trolley, sink and cupboard.
5. Large and well-defined activity zones.
6. Well-balanced decoration and display.
7. Use of some strong colours for the floor, wall and blinds.

Another three distinctive features also stand out in this classroom. Again, these were not prioritised by the statistical analysis, but could be expected to enhance the physical environment:

8. Pleasant views outside through the window.
9. Large sizes of window openings facilitating cross-ventilation.
10. Easy access to the outside playground.

It can be seen that the differences are quite subtle, but the impacts on learning are large.

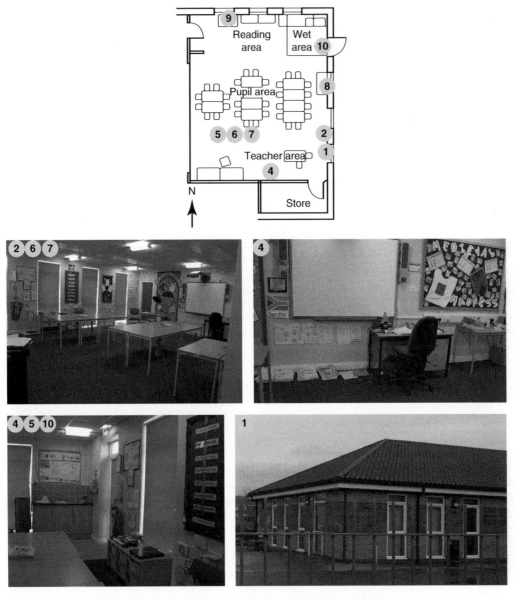

Figure 2.6   Classroom environment that has the highest predicted test improvement.

2.6   Conclusions*

The early sections of this chapter illustrated the complex and dynamic ways in which humans receive information about their surroundings through their senses. It was also shown that these multi-dimensional inputs can have profound but fragile and interactive effects on human health, mood and performance. Some basic neuroscience theories about the way the human brain makes calculations to link perceptions to actions were then used to cast light on a strategic framing of these impacts. This led to the proposition that design that aims to make the most of users' sensory response to spaces should take into account the issues of naturalness, individualisation and level of stimulation.

Core to this chapter is the view that an approach that focuses on the user's sensory perception of spaces opens up a new opportunity to optimize designs around user outcomes. This is not instead of the usual aesthetic aspirations of designers, but alongside this endeavour. This is an argument for an iterative 'inside-out-inside' design process, where the spaces making up the building/area are designed to meet the human needs of those pursuing activities in each space, then the building as a whole is compiled, then the functionality of each space is checked again, and so on. Of course this calls for a design effort that accommodates more information inputs and as such it is a significant design challenge. However, the benefits that could flow are enormous, e.g. schools in which pupils' well-being is improved and where they achieve enhanced academic results; hospitals in which patients need less pain relief and get better faster; offices in which workers are more efficient and productive; and homes and urban areas in which people can live more contently.

A web-based forum of over 450 corporate real estate and facilities management (CRE&FM) professionals (Varcoe, 2012) confirms a collective view as to the importance of these issues. Just to quote a few headlines: 'CRE&FM currently has no objective way of evidencing the value it brings to business … blocker to CRE&FM being seen as a business asset rather than a cost … the worker and their needs (effectiveness) are now the primary CRE&FM focus, not the workplace (efficiency) …' (p. 5).

As this chapter indicates, there seems to be tremendous headroom to move towards excellence, i.e. to provide safe, comfortable, sustainable *and stimulating* environments as called for in the European Construction Technology Platform Strategic Research Agenda (ECTP, 2005). It is suggested here that the pursuit of *optimal* built spaces for human use can productively use insights from neuroscience to gain a greater understanding of human responses in complex sensory situations. A wide range of factors appears to be important, such as the type of stimulus involved, its intensity or quality and factors such as the different reinforcement contingencies and the emotions associated with them. Such complexity reveals the difficulty facing laboratory work in this area. However, greater understanding of the relationship between the various stimuli and the subsequent responses could be gained, linking findings from laboratory work regarding light, colour, etc., with findings from case studies/experimental buildings involving different uses, such as housing, offices, etc., so that triangulation between the different results can lead to guiding conceptual and virtual models (Barrett and Barrett, 2003). As an example, evidence has been given from a research study of school design features that are strongly linked to enhanced learning progression amongst pupils.

The material brought together highlights the fact that finding optimal solutions is not a simple search for the answer. It is, rather, a subtle process of addressing multiple aspects, linking together the specialist knowledge that is available and moving towards a better understanding of the issues we need to address for specific users. This will help to create an informed context within which robust, multi-faceted design solutions can more confidently be derived. If successful, this will lead to the creation of spaces that provide people with sensory environments that help them reach, or sustain, their fullest possible potential. The three themes of naturalness, individualisation and level of stimulation have been suggested as productive foci to drive this work forward through a rich programme of multi-disciplinary research.

Therefore there exists an important challenge around the issue of better understanding, and evidencing, the holistic impact of spaces on users. Using the paradigm of the brain's response to multiple sensory inputs, there is initial evidence that causality between the complex characteristics of built environments and users' responses can be revealed. This sort of approach needs to be taken forward more broadly, and then powerful, evidence-based insights can be expected to inform the improved design, adaptation and use of built spaces. More than this, those in the facilities management function will be able to proactively deliver their service so that it has a maximum positive impact on the health, well-being and productivity of those in the core businesses supported. Achieving this orientation is the only way FM can go beyond cost containment and sustainably perform a positive strategic role.

Finally, it should be noted that facilities managers can use the sort of ideas set out above, but they also have unique access to data that can inform our knowledge of how users respond to different environments. One aspect of this is through evaluations of users' needs, which is the focus of the next chapter.

Endnote

This project has been supported from several directions. Much of the work in general started within the Salford Centre for Research and Innovation in the Built Environment (SCRI–EPSRC grant ref. EP/E001882/1). Nightingale Associates funded more focused work and facilitated the link to Blackpool Council. EPSRC has funded the main HEAD project (grant ref. EP/J015709/1) and this is the vehicle through which this body of work has been brought to this point.

References

Achilles, C., Finn, J. and Bain, H. (1998) Using class size to reduce the equity gap. *Educational Leadership*, **55**(4), 40–45.

Alexander, R. (1977) *Policy and Practice in Primary Education: Local Initiative*. Routledge, London.

Barrett, P.S. (2008) *Revaluing Construction*. Blackwell Publishing, Oxford.

Barrett, P.S. and Barrett, L.C. (2003) Research as a kaleidoscope on practice. *Construction Management and Economics*, **21**: 755–766.

Barrett, P.S., Zhang, Y., Moffat, J. *et al.* (2013) An holistic, multi-level analysis identifying the impact of classroom design on pupils' learning. *Building and Environment*, **59**, 678-689.

Begemann, S., Van den Beld, G. and Tenner, A. (1996) Daylight, artificial light and people in an office environment, overview of visual and biological responses. *Industrial Ergonomics*, **20**(3), 231–239.

Berglund, B., Bluyssen, P.C.G. *et al.* (1999) Sensory Evaluation of Indoor Air Quality: Report 20. *Environment and Quality of Life*, European Commission.

Bonnefoy, X. *et al.* (2004) *Review of Evidence on Housing and Health*. World Health Organisation, Budapest.

Bordass, B. and Leaman, A. (2005). Making feedback and post-occupancy evaluation routine. *Building Research and Information*, **33**(4), 361–375.

Boyce, P., Hunter, C. and Howlett, O. (2003) *The Benefits of Daylight Through Windows*. Lighting Research Centre, Rensselaer Polytechnic Institute, Troy, NY. pp. 1–88.

Carter, R. (1998) *Mapping the Mind*. Phoenix, London.

Christoffersen, J., Johnsen, K., Petersen, E. *et al.* (2000) *Windows and Daylight: A Post-occupancy Evaluation of Danish Offices*. CIBSE/ILE Joint Conference, University of York, CIBSE/ILE.

CIB Task Group 42 (2004) *Performance Criteria of Buildings for Health and Comfort* (ed. J. Sateri). International Council for Research and Innovation in Building and Construction (CIB), Rotterdam. pp. 3–70.

Coley, D.A. and Greeves, R. (2004) *The Effect of Low Ventilation Rates on the Cognitive Function of a Primary School Class*. Exeter University, Exeter.

Derval, D. (2010) *The Right Sensory Mix*. Springer, Heidelberg.

Durao, M.J. (2000) *Colour and Space*. Des Livro, Lisbon.

Eberhard, J.P. (2007) *Architecture and the Brain: A New Knowledge Base from Neuroscience*. Ostberg, Atlanta, GA.

ECTP (2005) Strategic Research Agenda for the European Construction Sector: Achieving a Sustainable and Competitive Construction Sector by 2030. ECTP, Paris.

Engelbrecht, K. (2003) The Impact of Color on Learning. Available from: http://www.coe.uga.edu/sdp1.

European Commission (EC) (2003) Ventilation, Good Indoor Air Quality and Rational Use of Energy. *Environment and Quality of Life*, Joint Research Centre–Institute for Health and Consumer Protection, Physical and Chemical Exposure Unit.

Evans, G. and Lepore, S. (1993) Non-auditory effects of noise on children: a critical review. *Children's Environments*, **10**(1), 43–72.

Figueiro, M., Rea, M. and Eggleston, G. (2003) Light therapy and Alzheimer's disease. *Sleep Review* (January–February).

Fisk, W.J. (2000). Health productivity gains from better indoor environments and their relationship with building energy efficiency. *Annual Review of Energy and the Environment*, **25**(1), 537–566.

Fjortoft, I. (2004) Landscape as playscape: the effects of natural environments on children's play and motor development. *Children, Youth and Environments*, **14**(2), 21–44.

Fjortoft, I. and Sageie, J. (1999) The natural environment as a playground for children: landscape description and analyses of a natural playscape. *Landscape and Urban Planning*, **48**, 83–97.

Frank, K.A. and Lepori, R.B. (2007). *Architecture from the Inside Out*. John Wiley & Sons, Inc., Hoboken, NJ.

Goswami, U. (2004) Neuroscience and education. *British Journal of Educational Psychology*, **74**(March), 1–14.

Gunnarsen, L. and Santos, A. (1999) Indoor Climate Optimization with Limited Resources. Danish Building Research Institute (SBi), Horsholm.

Hartig, T., Evans, G., Jarmner, L. *et al.* (2003) Tracking restoration in natural and urban field settings. *Journal of Environmental Psychology*, **23**, 109–123.

Hartley, H., Trinkler, I. and Burgess, N. (2003) Geometric determinants of human spatial memory. *Cognition*, **94**, 39–75.

Haynes, B. (2011) The impact of generational differences on the workplace. *Journal of Corporate Real Estate*, **13**(2), 98–108.

Heinrich, D.P. (1980) *Colour Helps Sell: Colour Theory and Colour Psychology for Commerce and Advertising*. Muster-Schmidt Verlag, Gottingen.

Heinrich, D.P. (1992) *The Law of Colour*. Muster-Schmidt Verlag, Gottingen.

Heschong Mahone Group (1999) Daylighting in Schools. Pacific Gas and Electric Company, Fair Oaks, CA.

Heschong Mahone Group (2003). *Windows and Classrooms*: A Study of Student Performance and the Indoor Environment. Californian Energy Commission, Fair Oaks, CA.

Hidaka, T. and Nishihara, N. (2004) Objective evaluation of chamber-music halls in Europe and Japan. *Journal of the Acoustical Society of America*, **116**(1), 357–372.

Higgins, S., Hall, E., Wall, K. *et al.* (2005) The Impact of School Environments: A Literature Review. Design Council, London.

Hund, A. and Plumert, J. (2004) The stability and flexibility of spatial categories. *Cognitive Psychology*, **50**, 1–44.

Hygge, S. (2003) Classroom experiments on the effects of different noise sources and sound levels on long-term recall and recognition in children. *Applied Cognitive Psychology*, **17**, 895–914.

Jeon, J.Y. and Barron, M. (2005) Evaluation of stage acoustics in Seoul Arts Center Concert Hall by measuring stage support. *Journal of the Acoustical Society of America*, **117**(1), 232–239.

Joseph, A. (2006) The Impact of Light on Outcomes in Health Care Settings. The Center for Health Design, Concord, CA.

Kaplan, S. (1995) The restorative benefits of nature: toward an integrative framework. *Environmental Psychology*, **15**, 169–182.

Leaman, A. and Bordass, B. (1999) Productivity in buildings: the 'killer' variables. *Building Research and Information*, **27**(1), 4–19.

Lehman, M.L. (2011) How sensory design brings value to buildings and their occupants. *Intelligent Buildings International*, **3**(1), 46–54.

Lindholm, G. (1995) Schoolyards: the significance of place properties to outdoor activities in schools. *Environment and Behaviour*, **27**, 259–293.

Mahnke, F. (1996) *Color, Environment and Human Response*. John Wiley & Sons, Inc., New York.

Marx, A., Fuhrer, U. and Hartig, T. (1999) The effects of classrooom seating arrangments on children's question-asking. *Learning Environments Research*, **2**, 249–263.

Moore, G. and Lackney, J. (1994) Educational Facilities for the Twenty-First Century: Research Analysis and Design Patterns, Report R94-91. School of Architecture and Urban Planning and University of Wisconsin–Milwaukee Center for Architecture and Urban Planning Research.

Nasar, J. (1999) *Design by Competition: Making Design Competition Work*. Cambridge University Press, Cambridge.

National Platform for the Built Environment (2008) Research Priorities for the UK Built Environment. Constructing Excellence, London.

Norman, D.A. (1998) *The Design of Everyday Things*. MIT Press, London.

Nuhfer, E. (1994) Some Aspects of an Ideal Classroom: Color, Carpet, Light, and Furniture.

Pallasmaa, J. (2009) *The Thinking Hand*. John Wiley & Sons, Ltd, Chichester.

Pernao, J. (2010). *The 'Otherness' of White. Colour and Light in Architecture*. IUAV, Knemesi, Venice.

Preiser, W. and Vischer, J.C. (2005). *Assessing Building Performance*. Elsevier-Butterworth-Heinemann, Oxford.

Quayle, M. and van der Lieck (1997) Growing community : a case for hybrid design. *Landscape and Planning*, **39**(2), 99–107.

Rea, M., Figueiro, M. and Bullough, J. (2002) Circadian photobiology: an emerging framework for lighting practice and research. *Lighting Research and Technology*, **34**(3), 177–187.

Rivlin, L. and Wolfe, M. (1972) The early history of a psychiatric hospital for children: expectations and reality. *Environment and Behaviour*, **4**(33), 33–73.

Rolls, E.T. (2007) *Emotion Explained*. Oxford University Press, Oxford.

Santos, A. and Gunnarsen, L. (1999) Indoor Climate Optimization with Limited Resources. SBi Report 314. Hoersholm, Denmark.

Saxon, R. (2006) Be Valuable: A Guide to Constructing Excellence in the Built Environment. Constructing Excellence, London.

Sundstrom, E., Bell, P., Busby, P. *et al.* (1996) Environmental psychology. *Annual Review Psychology*, **47**(1), 485–512.

Tanner, C. (2000) The influence of school design on academic achievement. *Journal of Educational Administration*, **38**(4), 309–330.

Ulrich, R. (1984) View through a window may influence recovery from surgery. *Science*, **224**, 420–421.

Ulrich, R. (1991) The effects of interior design on wellness: theory and recent scientific research. *Journal of Health Care Interior Design*, **3**(1), 97–109.

Ulrich, R., Quan, X., Zimring, C. *et al.* (2004) The Role of the Physical Environment in the Hospital of the 21st Century: A Once-in-a-Lifetime Opportunity. Designing the 21 st Century Hospital Project. The Center for Health Design, Texas A&M University, Concord.

Vischer, J.C. (2005) *Space Meets Status: Designing Workplace Performance*. Routledge, Abingdon.

Wohlwill, J. and Van Vliet, W. (1985) *Habitats for Children: The Impacts of Density*. Lawrence Erlbraum Associates, Hillsdale, NJ.

Wolfe, J., Kluender, K., Levi, D. *et al.* (2006) *Sensation and Perception*. Sinauer Associates, Sunderland, MA.

World Health Organisation (WHO) (2007) Large Analysis and Review of European Housing and Health Status. Copenhagen.

Zeisel, J. (2006) *Enquiry by Design*. W.W. Norton and Co., New York.

Zeisel, J., Silverstein, N., Hyde, J. *et al.* (2003) Environmental correlates to behavioral health outcomes in Alzheimer's special care units. *The Gerontologist*, **43**(5), 697–711.

3 Engaging with Stakeholder Needs

3.1 Introduction

3.1.1 Aims

It is widely accepted that the efficiency and effectiveness of an organisation is influenced by the physical environment in which it operates. Chapter 2 has argued that this is a fruitful focus for pursuing excellence in facilities management. A clear understanding of the workplace and its impact upon user behaviour is vital if opportunities for performance enhancement are to be exploited. This chapter endeavours to present 'user needs evaluation' as an analytical and structured approach to achieving this necessary level of understanding and to provide the platform for facilities managers to make their contribution to organisational goal achievement.

3.1.2 Context

At present, many organisations implement a linear building process, as shown in Figure 3.1 (Preiser, Vischer and White, 1991). Organisations will identify their need to build and will then work through the process from the planning stage to the occupancy stage. This same process will be repeated for every new building project that an organisation may undertake. Even though this is the typical method, it is not necessarily the best one. This chapter argues that organisations should instead implement a new building method, as shown in Figure 3.2.

The new method is cyclical rather than linear. Even though five of the stages are the same there is one important addition: the stage of evaluation. The latter is added, as organisations are often not making use of a valuable resource that they already have at their fingertips, namely their staff. Very few organisations ask their staff whether a building meets their requirements, even though the people that understand a building best are the people that use it every day. The cyclical method encourages organisations to learn from their staff whether a building is performing as well as it should. This information can be used in various ways: it can be fed forward into the design of a new building or it can be fed back to improve an existing building.

Facilities Management: The Dynamics of Excellence, Third Edition. Peter Barrett and Edward Finch.
© 2014 John Wiley & Sons, Ltd. Published 2014 by John Wiley & Sons, Ltd.

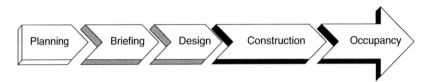

Figure 3.1 Traditional building process.

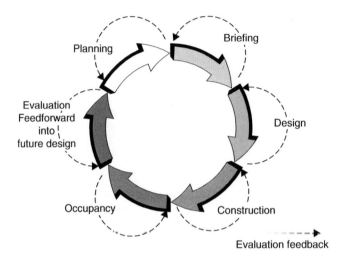

Figure 3.2 Cyclical building process.

Evaluation is the missing link in the design process. Evaluation, programming (briefing), and design are three linked activities drawing information from a systematic look at how people use existing environments. Analysing environments leads to programming (briefing) (Sanoff, 1968).

This statement makes it clear that evaluation and briefing are closely linked. Ongoing building evaluation, based on user intelligence, is the main concern of this chapter. Chapter 6 picks up on the special case of briefing for a new project.

3.1.3 Summary of the different sections

- Section 3.1. Introduction.
- Section 3.2. This introduces the subject of building evaluation and discusses how organisations can benefit by implementing appraisals on a regular basis. The importance of user knowledge is also demonstrated.
- Section 3.3. This presents three different post-occupancy evaluation techniques that could easily be used by facilities managers.
- Section 3.4. This considers various data collection methods that can be used during briefing and post-occupancy evaluations. Techniques for data analysis and presentation are also discussed briefly.

> **Case Study: Background information**
>
> The case study material in this chapter is derived from the study of one organisation. The organisation in question is involved with private healthcare and owns over 30 hospitals situated around Britain. Facilities management exists at four levels within the organisation: board level, corporate level, regional level and hospital level. In broad terms, the director of facilities and the corporate facilities group are responsible for setting down facilities management policy, while the regional groups coordinate the day-to-day facilities management activities, which are carried out at the different hospitals within their region.
>
> The case study elements are used to demonstrate how this particular organisation approaches the practice of building evaluation. Although the organisation is well-experienced in briefing terms, it has only recently become involved with building evaluations. Therefore, the relevant case studies demonstrate why the organisation became involved in building evaluations and how it carried them out.

3.2 The relationship between the facilities management function and user needs evaluation

Meeting user needs determines the degree to which the facilities management function contributes to the corporate success of the organisation or enterprise. A US facilities consultant has identified the 'Five Pillars of Quality' (Teicholz, 2001), which facilities functions need to be built upon:

- The customer drives the process.
- The facilities manager must be committed to continual improvement.
- Benchmarking and metrics are essential.
- Front-line workers must be empowered and held responsible.
- Marketing cannot be done by sitting in an office.

The more proactive facilities organisations have demonstrated the capability to go beyond the traditional means of identifying customer or user needs and sought to achieve a clearer understanding of expectations held by customers as to the quality and range of services provided. Customers have a set of predetermined expectations of service quality and a perception of what is being delivered. Where expectations exceed the perceived service, a service–quality gap appears and then customer dissatisfaction will be the outcome (see Chapter 9 for more on this perspective). In forming their service expectations, users will employ a number of criteria to establish the purpose of the service to be provided: how necessary it is seen to be; its relative importance compared to other services; the anticipated impact of the service provision; the costs likely to arise; and the level of potential failure associated with the activity. Furthermore, user expectations, and the subsequent evaluation of performance, are likely to be influenced by the anticipation of what could be delivered, the significance attributed to previous experiences (even if only tangentially relevant) and the ability to make comparisons with other service outcomes and feedback from the experiences of others, particularly if recent and local.

3.2.1   Aims

Facilities managers should:

- understand how building evaluations can contribute to organisational effectiveness;
- be able to communicate the importance of building evaluations to other people within an organisation.

3.2.2   The importance of the design of buildings

The design of a building is very important, as it has the power to affect how well an organisation can perform its function. At its most fundamental level, a building provides its users with basic elements that maintain the necessary comfort that allows them to carry out their jobs. These elements are shelter, light, heat and sanitation. In addition, facilities provide spatial arrangements that can aid specific activities and hinder others. Ease of movement, ease of communication and privacy are all dependent upon the design of a building. Therefore, if a building is designed without these basic requirements in mind, it is unlikely to provide a suitable working environment.

There is more to the design of working spaces than the provision of these essentials. Constant changes in office technology, business climates and human values affect many things, from organisational structures to the way people work. Consequently new managerial strategies are required to cope with these changes. To enable these strategic changes to take place, it will be necessary to rethink the way in which workspaces are designed. Workers now require far more from their workspaces than they have ever done before. This is due to the fact that workers are now better educated, have more job mobility and are more technically oriented. They are now concerned with the quality of work life and this includes the quality of their working environment.

Environmental planning was relatively simple in the past. It was based on the requirements of the organisation and management, not the requirements of the general employees. Buildings were designed to allow tasks to be carried out efficiently, with little concern for the comfort or needs of the workers. Organisations that consulted their staff about their preferences were few and far between.

In contrast, planning for today's ever-changing business operations requires organisations to think about the preferences of their workers, as well as their organisational objectives.

> Corporate environmental design must include an in-depth analysis of the corporation as a social organisation, the behaviour of the individuals within it, and the physical setting that houses it – a total system of interrelated parts (Moleski and Lang, 1986).

3.2.3   *Value of user knowledge/involvement*

Users may be pressurising managers to take their design requirements into consideration. However, organisations should realise that there are definite benefits to be gained from employee participation. They fall into two basic categories:

- organisational development;
- improvements to the built environment.

Organisational development

Participation encourages users to make decisions about their own environment. Employees realise that their views are important and this encourages feelings of personal responsibility; hence they become more motivated and committed to their jobs. If users are involved in the design of a facility, this can contribute to their understanding and acceptance of the final design. Alternatively, a lack of staff consultation may encourage feelings of rejection or hostility towards the new environment and the managers responsible for introducing it.

Improvement of the built environment

Users obviously come to know about the buildings that they work in; they are the experts on how well a facility works physically and operationally. Therefore, it makes sense to consult them when either refurbishing facilities or designing new ones. When considering existing buildings, users can pinpoint areas that do not assist them in carrying out their work, enabling them to be corrected. Similarly, when thinking about a new facility, users can save an organisation money by identifying designs that have created problems in the past.

3.2.4   *The importance of building appraisals for organisations*

How do organisations know if their facilities are supporting organisational goals and user requirements? The key is to introduce regular building appraisals. However, in most organisations, building appraisal methods are not very well developed. Organisations tend to have far more information on items such as photocopiers than they do on their buildings. Organisations that are relatively good at managing the rest of their assets often have very little information concerning the performance of their buildings. Those that possess data on areas such as energy costs could well have no information on how energy performance relates to employee comfort (Tom, 2008). Even if organisations have such information, it is unlikely that they will have tried to relate their present needs to what they are likely to require in a few years time.

The realisation that facilities may affect an organisation's effectiveness and its employees' welfare makes it imperative that building appraisals are introduced on a regular basis.

Facilities represent a new and untapped frontier for improving organisations' performance (Becker, 1990).

3.2.5   Facilities management and building appraisals

Logically, the department that is most suited to carrying out building appraisal is the one that is responsible for an organisation's buildings: the facilities management department. All service groups within organisations today are being asked to demonstrate how their activities are helping the organisation to achieve its business objectives. Facilities represent a substantial percentage of most organisations' assets and also a substantial proportion of their operating costs; thus facilities managers have to justify why and how money is being spent.

When the facilities management unit lacks reliable and comparable data on building performance and costs, its ability to make its most basic decisions is impaired, as is its ability to make a convincing case for its recommendations. The ability to demonstrate the facilities management unit's organisational effectiveness is hampered without such information. Reporting to management is easier and more convincing when the consequences of decisions can be demonstrated. The facilities management unit should be able to show better, for example, that the new planning processes, procedures or space guidelines have lowered the cost or the number of renovations and have enabled the building to accommodate organisational change, a new management style or dramatic shifts in group size.

3.2.6   Uses and benefits of building appraisals

Building appraisals provide an opportunity for an organisation to see how well a particular facility meets their requirements from various viewpoints. In the most general terms, appraisals can serve two purposes:

- to improve the current situation, known as post-occupancy evaluation;
- to aid in the design of future buildings, known as briefing.

The first of these two areas is covered in detail in Section 3.3 and the second in Chapter 6, but the following paragraphs give an idea of just some of the more specific issues to which appraisals can be applied.

Appraisals can be valuable when organisations are either shrinking or expanding and when they are renovating or when building new. When deciding whether to lease or purchase new facilities, performance and cost data are invaluable. They can be used as a preliminary form of architectural briefing, to guide the search for design solutions. It is difficult to decide what is needed in a new building unless an organisation has assessed what it has at present.

For long-range strategic planning, building appraisal provides information about what kinds of buildings will be needed in the future to accommodate the organisation's expected development. Knowledge of which buildings are performing poorly and which buildings are performing well is important for consideration of long-term strategy.

Decisions about managing the occupancy of the building stock require comparable and reliable data. Which buildings or which areas within the current stock of buildings have the best location for a particular unit? Which building will best meet the needs of that unit over the next several years so that disruption and cost incurred by frequent relocations or renovations can be minimised?

Operational and maintenance decisions can also benefit from building performance data. Which types of buildings or which elements within buildings require the least maintenance, are the most energy efficient and incur the fewest breakdowns and repairs? Which are the easiest to clean? What cleaning or maintenance strategies work best for particular buildings?

The above examples demonstrate the various ways that building appraisal can be used. It can be applied to existing buildings, proposed designs, cost or occupant satisfaction.

Case Study: Evaluation benefits

The organisation had been involved with many building projects over the years, both refurbishment and new-build. The organisation was typical, in that once building work was completed project managers would turn their attention to their next project. If problems occurred after occupation then a hospital manager would contact the project manager and the problem would be corrected as far as possible. No formal procedure existed for recording these problems; hence the different project managers were making similar mistakes on other projects. Thus the organisation was following a linear building programme, rather than a cyclical one where the organisation could learn from its mistakes.

Through informal discussions within the facilities department it became apparent that the lack of feedback could prove costly. If problems were not identified quickly there was the possibility that they could be repeated throughout the other 30 hospitals. Therefore, it was decided that a formal evaluation programme would be initiated so that mistakes could be identified and recorded to prevent the same thing happening again. In addition, the evaluations would be used to highlight successful designs so that these could be used again. The facilities manager suggested that the information collected during the evaluations could be used to compile guidelines for an ideal hospital, against which all future new-build and refurbishment projects could be measured.

Comment
The organisation initially decided to implement an evaluation programme to prevent costly mistakes from being repeated in other hospitals. However, the facilities manager realised that evaluations could be used to identify good points as well as bad, thus demonstrating the versatility of evaluations.

3.3   Post-occupancy evaluation (POE)

3.3.1   *Aims*

Facilities managers should:

- understand the potential benefits that an organisation can gain through the use of building evaluations;
- be able to learn the necessary skills to enable them to conduct their own building evaluations.

3.3.2   *Building evaluation systems*

Various methods for building evaluations exist, but they can broadly be divided into two categories: user-based systems or expert-based systems. The first system uses a building's occupants to evaluate the suitability of a building for their particular needs and hence is also known as post-occupancy evaluation (POE). The second method relies on experts' assessments and typically covers far more areas, such as provision for information technology, organisational growth, changes in staff work style and energy efficiency.

As this book is directed towards good practice in facilities management, only POE methods are described in this section, as these can be carried out by the facilities management department, whereas expert assessments cannot. However, facilities managers should recognise that expert assistance may be necessary if post-occupancy evaluations highlight problems that are outside the capabilities of the organization (Preiser, Rabinowitz and White, 1988). For example, various expert systems have been developed and a selection are described in *The Total Workplace* (Becker, 1990).

3.3.3   *POE methods*

Users of buildings often complain that their workplace is not designed to meet all of their work needs. Facilities are designed by professionals who believe that they understand how people use buildings. Unfortunately this is rarely the case and issues that are important to users are often overlooked by designers. In post-occupancy evaluation methods, the focus is on user satisfaction.

At its most basic level, post-occupancy evaluation is a formal evaluation of a building by its occupants after it has been completed in order to identify areas that do not meet users' requirements. However, despite its title, post-occupancy evaluation is also a useful tool when planning new facilities, as data generated during an evaluation can be used in the briefing process for a new building (see Chapter 6).

Potential benefits arising from the use of post-occupancy evaluations range from short term through to long term, as Table 3.1 illustrates (Preiser, Rabinowitz and White, 1988).

Various methods for POE have been developed, but the three examples that follow have been selected because they demonstrate the different techniques and

Table 3.1   Benefits of POE.

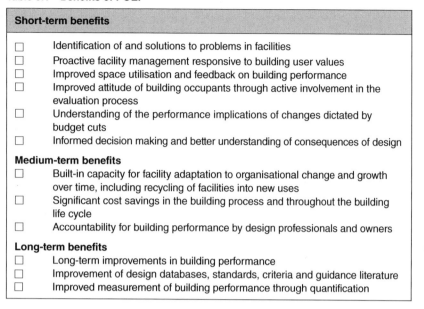

Short-term benefits
☐ Identification of and solutions to problems in facilities
☐ Proactive facility management responsive to building user values
☐ Improved space utilisation and feedback on building performance
☐ Improved attitude of building occupants through active involvement in the evaluation process
☐ Understanding of the performance implications of changes dictated by budget cuts
☐ Informed decision making and better understanding of consequences of design

Medium-term benefits
- ☐ Built-in capacity for facility adaptation to organisational change and growth over time, including recycling of facilities into new uses
- ☐ Significant cost savings in the building process and throughout the building life cycle
- ☐ Accountability for building performance by design professionals and owners

Long-term benefits
- ☐ Long-term improvements in building performance
- ☐ Improvement of design databases, standards, criteria and guidance literature
- ☐ Improved measurement of building performance through quantification

uses of POE. It is hoped that these examples will provide the facilities managers with enough information to enable them to conduct their own POEs:

- partial user participation;
- full user participation;
- management POE.

Partial user participation

The following model was developed by Preiser, Rabinowitz and White (1988). It is an example of a POE whereby users are only partially involved in the evaluation process. Experienced evaluators conduct the process and users only participate at the request of these evaluators.

Three levels of effort are proposed in this model (Figure 3.3). The process selected depends upon finances, time, manpower and the required outcome. However, each level contains the same procedures of planning, conducting and applying.

Level 1: indicative POE
This POE provides an indication of major successes and failures of a building's overall performance. It is normally carried out by an experienced evaluator, who should ideally be familiar with the building type being evaluated, and is completed in a very short time span. Thus, data collection needs to be quick and easily accessible. Methods used include: archival document evaluation, walk-through evaluations and interviews with staff. The findings are usually presented in the form of a short report, outlining the purpose of the evaluation, the data collection methods used, findings and recommendations.

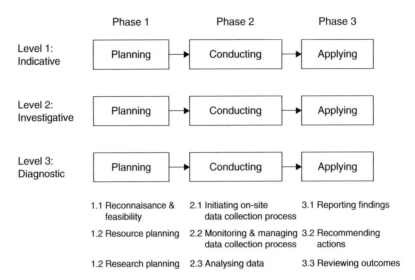

1.1 Reconnaisance & feasibility
2.1 Initiating on-site data collection process
3.1 Reporting findings

1.2 Resource planning
2.2 Monitoring & managing data collection process
3.2 Recommending actions

1.2 Research planning
2.3 Analysing data
3.3 Reviewing outcomes

Figure 3.3   Preiser's POE process model.

Level 2: investigative POE

This is often instigated as a result of a problem identified during an indicative POE. It is likely that before a solution can be proposed, the problem needs to be studied in more detail. Unlike the indicative method, whereby the evaluators make judgements themselves due to a lack of time, this system relies on more sophisticated data collection methods to produce results. Initially the evaluators undertake state-of-the-art literature reviews and study recent, similar, facilities. Then comparisons are made between these and the building being assessed, to see why problems may have occurred and to identify possible solutions. The findings are normally presented in a report, which identifies the specific problems studied and proposes recommendations for action. Annotated plans and photographs may be used to clarify findings.

Level 3: diagnostic POE

This type of POE aims to improve not only the particular facility being evaluated but also to influence the future design of similar facilities. Typically it will follow a multi-method strategy, including questionnaires, surveys, observations and physical measurements, all of which will allow comparisons to be made with other facilities. A diagnostic POE is likely to take several months at a minimum to complete. The results drawn from such research are long-term oriented, relating not only to the improvement of a particular facility but also to improvement of a specific building type.

Full user participation

In this type of POE, users are fully involved throughout the evaluation. People experienced in the process of evaluation are still involved, but their function is purely to guide the participants through the process rather than to make judgements. An example of this type of POE has been developed by Kernohan *et al.* (1992). Every evaluation will include the same three core events, which are described in more detail later, namely:

- introductory meeting;
- touring interview;
- review meeting.

It is also fundamental to the process that the following groups are involved with the evaluation:

- *Participant groups*, who evaluate the building. Participant groups represent the different interests in a building, both users and providers. The interests typically include those of occupants, visitors, owners, tenant organisations, makers, traders and maintainers. Representatives from each of these interests should be selected to form small groups that can participate in the evaluation. Each participant group is involved in the three-stage process and evaluates the building, taking into account the particular interests represented in the group.
- *Facilitators*, who assist participants to make their evaluations. Facilitators are there solely to support participants in their assessment. They do not evaluate the building or do any other kind of evaluation. Facilitators have a neutral role throughout. For the purpose of this book, it is assumed that members of the facilities management group will take on the role of the facilitators (see the next section on facilitation training). Both participants and facilitators may play a part in initiating evaluations and monitoring outcomes, but their prime activity is the evaluation itself. It is only participants and facilitators who are concerned with the on-site activities of the generic evaluation process.
- *Managers*, who authorise the evaluation. Managers are not normally concerned with the on-site activities, although they may be represented in a participant group. Their role is administrative and supportive. They may initiate, approve and authorise an evaluation and they have the responsibility for ensuring there is action on the outcomes and for the ongoing management of that action.

Training in facilitation skills

The skills needed to be able to facilitate an evaluation can be acquired through practice. However, facilitators need specific attitudinal and communication skills. They should be good at listening and be able to discard their own personal and professional attitudes during evaluations. They should be clear that it is the participant groups who evaluate the facility and not the facilitators.

Before conducting proper evaluations, it is a good idea to perform a test run within the facilities management department. Run through the three generic evaluation activities, but concentrate on a just a few rooms. Try to formulate a

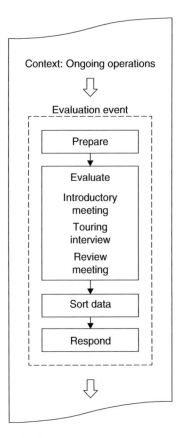

Figure 3.4   The generic evaluation process.

couple of recommendations from the issues that are raised during the tour. Going through this exercise will provide the facilitator with a good insight into what it is like to be a member of a participant group.

The generic evaluation process

The generic evaluation process is outlined in Figure 3.4. The three core stages mentioned previously are highlighted and will be conducted in the same manner for every evaluation. The other stages are likely to differ from evaluation to evaluation. Hence, the three core stages are now described, whilst the other stages will be looked at later.

1.   *Introductory meeting.* The facilitators meet with the participant group to explain the evaluation process and the procedures of the touring interview and review. Group members are encouraged to discuss their connection with the facility and raise topics that they feel are important. The route to be taken on the tour is then discussed, so that areas of concern can be visited. Each group does not have to follow the same route, as obviously different groups will be worried about different aspects.

Table 3.2   Facilitation tasks.

Preparation		tick √
Terms of reference	Check the purpose of the evaluation	☐
	Establish authorisation to carry out the evaluation	☐
	Check the lead time for planning the event	☐
	Estimate the costs, and who will pay	☐
	Ensure that there is money 'in the back pocket' for fine-tuning	☐
Preparing off-site	Gain a working knowledge of the facility and its occupants	☐
	Select the interest groups to take part in the evaluation	☐
	Allocate facilitators' roles	☐
	Prepare a workplan	☐
Preparing on-site	Meet senior manager(s) in the occupancy group(s)	☐
	Tour the facility for familiarisation purposes	☐
	Prepare meeting place	☐
Evaluation generic core	(Repeated with each evaluation group)	
Introductory meeting	Welcome participants and explain purpose and roles	☐
	Outline the touring interview process	☐
	Invite participants to ask questions and express their views	☐
	Decide which part of the facility to tour	☐
	Check that the participants know what will happen	☐
Touring interview	Conduct the tour as a conversation on the move, with prompts	☐
	Take a note of each topic raised during the tour	☐
	Display the list of topics before the review meeting starts	☐
Review meeting	Review the topics raised during the introduction and tour	☐
	Discuss the topics	☐
	Facilitate the process of preparing recommendations	☐
	Sort priorities among the recommendations	☐
Sort data and respond	Collate and classify information from the evaluation groups	☐
Climate for action	Document evaluation outcomes in a database, or report, or both	☐
	Facilitate a general review meeting (if asked to)	☐
	Clarify the means for achieving action	☐

2. *Touring interview.* Each participant group walks through the building with the facilitators, following the agreed route. Group members should discuss their views of the facility during the tour. The facilitator can use standard open-ended questions as prompts, but should be careful not to ask direct questions – in this type of evaluation the objective is to obtain users' views and not the views of the facilitator. Topics raised during the discussions are noted so that they can be discussed during the review meeting.

3. *Review meeting.* At this meeting, the different issues raised during the tour are discussed. It is helpful if the facilitator produces a record of the meeting, e.g. on a flip chart, so that it can be referred to later. The participant group should prioritise its concerns, so that its major problems can be looked at first; this is obviously critical if only limited finances are available.

Facilitation guidelines

The facilities manager should now understand how the three core stages are conducted. The other stages will vary depending upon the required output of the project. Table 3.2 summarises what activities may be included in the stages before and after the core. These activities are then described in detail in the following pages.

When preparing an evaluation the facilitator may find it useful to use Table 3.2 as a checklist to ensure that nothing has been omitted.

1. *Preparation.* It is important to ascertain why an evaluation is being conducted and on whose authority. Evaluations obviously need to be directed towards an outcome, so check what results are expected. Check time and budget allowances, who will pay for reports, etc. Try to ensure that some finances will be available on the completion of the evaluation, so that certain items can be dealt with immediately. If nothing changes as a result of an evaluation, both users and management will wonder why they agreed to participate.
2. *Selection of interest groups.* It is best to keep the size of participant groups small, between three to seven people, for practical reasons. People must be able to hear what is going on during the touring interview. Groups can be made from either single or mixed interests. However, the latter is not a good idea unless the facilitator is experienced, as there may be too many conflicting views or people may be afraid to express their views in front of people they do not know. Similarly, people may feel intimidated if their managers are present.
3. *Allocation of facilitators' roles.* Most evaluations should have at least two facilitators, one to guide the process and one to take notes. Decide beforehand who will do what. Will anybody take photographs or slides of relevant areas?
4. *Workplan preparation.* It is essential that a workplan is prepared before the evaluation. This should ensure that all aspects of the evaluation are planned for. Check that all participant groups are available on the date of the evaluation. Decide in which order groups are to be dealt with. Check that managers are informed of the agenda and appropriate times when their workers will be involved. Ensure that a room will be set aside for the meetings.
5. *Post-evaluation activities.* It is a good idea to produce a report following the evaluation, so that people can see that their recommendations have been recorded accurately. Do not try to include every comment that was made during the process. A typical report may consist of:

    (a) a cover sheet,
    (b) a contents list,
    (c) a summary of the evaluation method,
    (d) a brief background to the building project,
    (e) a photographic, drawn and written description of the building,
    (f) the main recommendations that were proposed during the review meetings,
    (g) photographs where appropriate to illustrate recommendations or particular problems.

Sometimes the distribution of a report may be enough to promote action. If this is the case it is important to make sure that all participants are kept informed of any agreed action. However, in most cases a general review meeting is held to agree on how the recommendations should be dealt with.

Before a general meeting is held it is a good idea to discuss with the appropriate managers their attitudes to the recommendations. There is no point in formulating proposals at a meeting if managers are likely to veto them at a later stage. It is also sensible to distribute copies of the report prior to a meeting, so that everyone understands what issues will be discussed.

6. *General review meeting.* The purpose of a general review meeting is to discuss and agree recommendations for action. Participants should be reminded of the various proposals made previously by each of the groups. It may be useful to display these on the wall, so that they can be addressed during the meeting.

    During the meeting, facilitators should help the participants to establish which recommendations have priority. Obviously different groups may feel that different issues are important, so it is necessary to ensure that one group does not dominate the meeting. The final outcome should be an agreed list of priorities. If possible, conclude with an explanation of what action will now be taken. Management may have already given authority for certain problems to be dealt with. If so, inform the participants so that they can see that management is committed to improving problem areas. Finally, the mechanism for keeping staff informed of further actions should be made clear.

POE as a management aid

It should be remembered that POEs not only serve to improve physical conditions but can also act as an aid to management. In POEs, the process of consultation can be as important as the information collected. POEs can be used to gain the trust and support of staff, so that when changes are necessary people may be more cooperative.

POEs can also serve to highlight general management or personnel problems. 'It is a psychological fact that people often unconsciously blame their visible or tangible surroundings for problems which have intangible or invisible causes' (Ellis, 1987). If a facilities manager takes environmental complaints at face value, he may spend a lot of money on physical solutions that are not going to solve the real underlying problems. A complaint about bad air, for example, could actually stem from dissatisfaction about overcrowding. In such a case, investment in a new air conditioning system would still not solve the real issue. If a facilities manager suspects that environmental

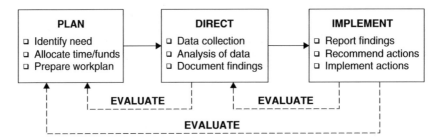

Figure 3.5   POE composite process diagram.

problems are not the real cause of complaints, then he should try to investigate further. A good technique for this is to hold a small group meeting, whereby those involved are encouraged to discuss their problems. Again the actual process of taking part may help to defuse the situation. If people are allowed to discuss their problems freely, it is likely that any personnel problems will become apparent. Obviously, if it appears that the environmental complaint is based on fact, it is up to the facilities manager to initiate any necessary actions.

3.3.4   Summary

1. POEs can be used for different purposes. It is up to the facilities manager to identify the purpose of an evaluation before selecting a suitable method. Is the POE to be used to:
   - Improve the physical surroundings?
   - Gain the support of staff through participation?
   - Collect information to be fed into future building designs?
2. If staff are to be involved in a POE, the facilities manager should ensure that there are visible results or else staff will lose confidence in the facilities department.
3. Before conducting a POE, the facilities manager should check that all the different stages have been planned for. Figure 3.4 can be used as a reminder of the major stages, which are applicable to all of the POE methods described in this section.
4. The POE process, like the briefing process, should be regarded as cyclical, as shown in Figure 3.5. If, after any stage, information seems to be missing, then the facilities manager should be able to step back and collect the necessary information. In addition, after a POE has been completed, the facilities manager should review the project to see if any part of the process could be improved upon next time.

Case Study: POE

As stated earlier (Section 3.2.6, Case Study: Evaluation benefits), no formal feedback procedures were undertaken by the facilities department to see what users thought of new building work. Problems with newly completed projects were dealt with promptly if they were reported, but users' views were not actively sought after each project.

Last year, however, five major refurbishment projects were completed and so the facilities manager decided that it might be a good time to actually introduce a formal post-occupancy evaluation programme. Not only would this ensure that all problems were addressed but it would also enable the department to identify which design solutions were preferred by the users. Previously, even though the three project managers all worked in the same room, they each had their own way of doing things that the others did not necessarily agree with. The facilities manager suggested that the information arising from the evaluations could be used to compile guidelines for an ideal hospital, against which all future new-build and refurbishment projects could be measured, thus ensuring a more unified approach to specific situations.

So that similar information could be collected at each site, a standard approach was agreed upon. A questionnaire would be completed for each new/refurbished room by the appropriate departmental head in conjunction with a project manager. The former would act as a representative for the whole department and discuss any problems/good solutions before the evaluation exercise. Consequently, one of the project managers was assigned to design a post-occupancy evaluation questionnaire. Once he had compiled a nine-page form, it was suggested (by the author) that it would be sensible to conduct a pilot test to see if the document was easy to use and promoted useful responses. However, it was decided that this was an unnecessary waste of time and so the project manager went straight to one of the five hospitals to conduct the first evaluation.

It very quickly became apparent that the document was far too long, as it took over an hour to complete each room. In addition, many of the questions were unnecessary and users believed they had to try to find a fault with each listed component, even if there was nothing wrong. Thus if the same questionnaire had remained in use, it could have taken weeks to do just one hospital! So the project manager designed a new, three-page questionnaire in consultation with the author, which was intended to act as a prompt rather than having to complete every section (see the Appendix).

Even though the evaluations have not yet been completed, attention has already been drawn to a number of features that could be improved in future designs. In an operating theatre department, for example, it is essential that dust and dirt are kept to an absolute minimum. However, these requirements have been overlooked in a number of ways, including unnecessary skirting boards, pipework that has not been enclosed and corridor walls that are falling to pieces as they are too narrow for the easy movement of trolleys. Further problems have been encountered in an X-ray department, where several rooms were just too small to be used properly. However, perhaps the worst case was not providing a bathroom close enough for patients who had been given a barium meal!

Comment

The above example demonstrates the importance of the planning stage in any post-occupancy evaluation exercise. By not conducting a pilot test, the department wasted both time and resources redesigning the questionnaire.

The problems that have been identified above may seem relatively minor, but such mistakes could prove very costly to correct if they were repeated throughout the organisation's 30 hospitals.

3.4   Data collection: Methods, analysis and presentation

3.4.1   Aims

The facilities manager should:

- be aware of the various methods that exist for data collection;
- understand the benefits and drawbacks of the different collection methods;
- understand how to approach data analysis;
- know the various techniques available for the presentation of results.

3.4.2   Context

Data collection and analysis are obviously very important parts of both briefing and evaluation; without sufficient data it is difficult to make informed decisions. The techniques covered in this chapter can be used during either process.

3.4.3   Data collection methods

The different methods covered in this section are listed in Table 3.3. The associated benefits and drawbacks are described briefly so that facilities managers can quickly select the appropriate method for a particular situation. This list is by no means exhaustive, however, as these particular techniques have been selected as they are the most useful and well-known ways of collecting data (Zeisel, 1984).

Before embarking on any data collection programme, it is worth bearing in mind the following points:

Table 3.3   Data collection methods.

Standardised questionnaires	Benefits:	☐ Generating quantitative data
		☐ Quickly tapping a broad cross section of employees
		☐ Enabling a statistical analysis of subgroups
	Drawbacks:	☐ Probing responses
		☐ Understanding complex non-statistical relationships
		☐ Generating goodwill and confidence in the process
Focused interviews	Benefits:	☐ Probing responses
		☐ Engendering goodwill
		☐ Understanding complex relationships
	Drawbacks:	☐ Developing quantitative data
		☐ Quickly tapping a broad sample of employees
		☐ Time consuming and expensive
Structured observation	Benefits:	☐ Checking information given in surveys and interviews
		☐ (If systematic) generating quantitative data
		☐ Generating visual evidence to support interviews and surveys
		☐ Getting at issues that employees have difficulty verbalising
	Drawbacks:	☐ Understanding why something is occurring
		☐ Generating goodwill, unless coupled with interviews
Tracing	Benefits:	☐ Unobtrusive
		☐ Inexpensive data collection
	Drawback:	☐ Understanding why something is occurring
Literature search	Benefits:	☐ Eliciting responses about other buildings
		☐ Stimulating the imagination
	Drawbacks:	☐ Understanding how well a building functions
		☐ Time consuming
Study visit	Benefits:	☐ Eliciting responses about existing buildings
		☐ Stimulating the imagination
	Drawbacks:	☐ Background research time consuming
		☐ Understanding complex relationships
Archival records	Benefits:	☐ Unobtrusive
		☐ Inexpensive data collection
		☐ A check on other sources of information
	Drawbacks:	☐ An understanding of why something is happening
		☐ A detailed look at an issue
		☐ An accurate interpretation of data
Simulation	Benefits:	☐ Exploring 'what if' possibilities
		☐ Eliciting responses to new designs or plans
		☐ Removing scepticism that something will happen
		☐ Avoiding costly mistakes
		☐ Stimulating the imagination
		☐ Generating enthusiasm and excitement
	Drawbacks:	☐ Getting a completely realistic response

- Using multiple methods is likely to produce better results than using one single technique. For example, structured observation will only highlight what is occurring, rather than why.
- Information should not be collected just for the current situation. Organisations are forever changing; therefore possible future requirements should also be catered for.

Standardised questionnaires

Questionnaires are a very traditional method of obtaining data. They are often used to discover regularities among groups of people, by comparing answers to the same set of questions. Analysis of questionnaire responses can provide precise numerical data, from which tables, graphs, etc., can be produced.

Before writing a questionnaire, researchers should carry out preliminary investigations, such as focused interviews. This enables the researcher to establish what type of answers respondents will give to specific questions; people will not always react to a question in the way that was expected. Following this, a standardised questionnaire can be compiled. Once the questionnaire is written, it should be pre-tested to see how people react. This should draw out any potential problems and the questionnaire can be altered to take these into consideration. It is worth remembering that employees are busy people and will not want to waste their time filling in pages of answers; therefore questionnaires should be kept as short and as simple as possible. It is helpful to outline at the start what the aim of the questionnaire is; if people can see that it will benefit them they are more likely to complete the form.

Data produced by this method is good for establishing trends, but does not necessarily go deep enough to find out why things have occurred. Therefore, used together with observation methods and focused interviews, standardised questionnaires will help produce a fuller picture of the situation being studied.

Focused interviews

A focused interview can be used to establish in depth what individuals or groups think about a particular situation. Before carrying out an interview, the researcher should undertake some basic groundwork and should try to establish what topics are relevant to the situation. A questionnaire, for example, may have already been distributed, which highlighted several areas for concern. On the basis of this analysis, the researcher develops an 'interview guide', which lays down the topics that should be covered during an interview. Whilst conducting the interview the researcher can ask further questions to clarify points or to enlarge on specific issues. However, the interviewer should ensure that they do not influence the answers in any way; it is their job to keep the interview flowing without directing it.

Structured observation

There are several methods of structured observation also called direct observation. However, a technique that is both systematic and quantitative is known as *behavioural mapping*. This is when an observer records where and when certain behaviour occurs in a specific setting. Used over a day, week or month, such behaviour records allow the observer to build up a picture of which areas within a building are being used by what sort of people, in what ways and at what times. If, for example, a lounge at the end of a corridor is hardly used, compared to one in the middle of a circulation route, then a facilities manager could ask if this space could be used more productively for something else.

Tracing (unobtrusive observation)

Observing physical traces means systematically looking at physical surroundings to find reflections of previous activity not produced in order to be measured by researchers (Powell, 1991).

Traces may have been unconsciously left behind (e.g. paths across a field) or may be conscious changes people have made to their surroundings (e.g. a curtain hung over an open doorway). From such traces, researchers can ascertain how people actually use the environments that they work/live in. Facilities managers could use the technique to see how many changes staff make to their workspaces etc. in order for them to meet their particular needs. This method is unobtrusive and inexpensive. However, it also has its drawbacks, as without consultation with the users, researchers may make false assumptions. So again it should be used in conjunction with another technique. Methods of recording observations include: annotated diagrams, drawings, photographs and counting.

Traces can be divided into the following four groups:

- by-products of use;
- adaptations for use;
- displays of self;
- public messages.

Examples of these different traces are presented in Table 3.4.

Literature search

This method enables the client/designer to identify similar buildings and organisations, which may provide useful information on how other people have approached similar issues. The designer may use this method to ascertain the client's reactions to different architectural styles before any initial designing occurs.

Study visit

This method enables users, clients and designers to learn from the experiences of others. Study visits can be thought of as POEs of other peoples' facilities. Visits to similar organisations and buildings can make people aware of the different ways in which other people have designed buildings for the same use. Study visits may highlight problems with a particular building type or design solution, allowing designers and clients to avoid similar costly mistakes in the future. A building that seems impressive in a photograph will not necessarily function very well from its users' point of view. Obviously it will be impossible to visit too many buildings, so try to visit the ones that are the most comparable.

Archival records

This method is inexpensive, but again will only relate to what happened, rather than why. In this case, researchers consult records that have been collected by

Table 3.4 Physical traces.

By-products of use	These can be useful to the facilities manager to establish if people use spaces for the purposes they were initially designed for.
Erosions	Parts of the environment can be worn away indicating that an area has had more use than it was originally designed for. New routes may become apparent across a section of grass, indicating that the original design did not take into account how often people in one building would need to cross to another.
Leftovers	These are physical objects that may be left behind indicating how people have made use of a setting. Cigarette stubs, for example, left behind in a washroom may indicate that there is a need for a dedicated smoking room. Leftovers help to differentiate between places where planned activities have occurred and places where unplanned activities have taken place.
Missing traces	A lack of erosions or traces may help to identify areas which are being underused. A coffee/rest area without any empty cups or magazines etc. may demonstrate to the facilities manager that the space could be put to better use.
Adaptations for use	**When people find that their physical environment does not allow them to do something they want to do, they change their surroundings; they become designers. Adaptive traces are significant for facilities managers and designers because they demonstrate how people would choose to design their own environments if consulted.**
Props	New props are often added to a setting to allow for new activities. This may be due to a change in the function of a room or may be because certain activities were considered too expensive to allow for during the original design stage. Comfortable chairs and a low table may have been added to allow informal meetings to take place in someone's office, rather than having to occupy an official meeting room.
Separations	Changes may have been made to separate spaces that were previously together. A large open plan office, for example, may have been divided up by partitions to increase privacy.
Connections	Adaptations may have been made to allow for increased movement or communication between spaces that were designed to be physically separate. A door, for instance, between two offices may be permanently fixed ajar as the users work together as a team.
Displays of self	**People change their environments so that a place can be associated with them in particular. An environment which allows for no personalisation may result in workers that are unhappy with their surroundings, which may have a detrimental effect on how they view their organisation.**
Personalization	In work environments people often utilise space for their personal possessions, such as family photos or certificates. Facilities managers should bear this in mind when considering new designs.
Identification	People use their environments to enable others to identify them more easily. If employees have placed temporary name plates on partitions etc. it may suggest that a new design should provide fixed name plates as standard.
Public messages	**Physical environments can be used to communicate a specific message to the public at large.**
Official	How often does the name of an organisation appear around a building? How important is corporate image to the organisation? Are visitors prevented from entering certain areas by the use of *private* signs?
Unofficial	Are there a number of unofficial direction signs written on paper distributed around the building? If so, perhaps the official direction signs are inadequate.

the organisation as part of its regular record keeping. Medical records, staff turnover and rates of absenteeism could all be used to assess satisfaction with a building. If, for example, staff turnover is higher in one building than another, could the old HVAC system be responsible? Of all the methods considered this is probably the least reliable and should really only be used to verify the results from another technique.

Simulation

Simulation is not a method for initial data collection, but it can be a useful tool for obtaining reactions to new proposals. Simulation techniques include photographs, models, drawings, full-scale mock-ups, computer drawings, games and video animation. The method selected will depend upon the resources available. However, the cost of the simulation should be considered in relation to the cost of making a major mistake. A full-size mock-up of a new workstation may seem expensive, but it is better to get negative feedback at this stage, rather then when 40 new workstations have been installed that do not meet users' requirements.

3.4.4   Data analysis

The overall objective of data analysis is to interpret the collected information, so that useful recommendations can be made for existing or future buildings. Various methods for analysis of data exist. However, for most POEs and briefing projects simple techniques can produce useful results.

Data analysis cannot be left to the final stages of a project; it must be thought of early on in the process, as it will have implications throughout. Facilities managers should establish at the outset how final results are likely to be presented, as this will affect data collection methods and data analysis. If an organisation's managers, for example, like to have graphs to demonstrate results, then this suggests that questionnaires may be used, so that quantitative information can be manipulated to produce graphs. Budget and time schedules should also be considered to ensure that allowances for data analysis have been included.

Once data have been collected they can be analysed in various ways, either by hand or through the use of computers. At the most basic level, questionnaire responses can often be analysed simply by counting how many people answer in the same way. Alternatively, if the project team wants to do more sophisticated analysis, a simple statistical package can be employed. The choice of method depends upon the required output.

When interpreting results, try to identify areas where people agree and areas where people disagree, so that appropriate action can be initiated. If the majority of people in one building complain about the heat then there is likely to be a problem. However, if there is a certain amount of contradiction about a subject it is necessary to try to establish why contradictions exist. This may mean re-examining the data or performing further data collection. Once all of the data

---

Case Study: Use of multiple data collection methods

One of the office-based departments within the organisation felt that it might be able to reduce costs through a revised use of space. Hence an external space planning consultant was engaged to propose an alternative solution. Before any decisions could be made it was necessary to assess how the space was used at present; therefore the consultants conducted a post-occupancy evaluation.

During their evaluation, the consultants employed four different data collection methods:

- *Questionnaires.* All staff were asked to complete a short questionnaire, which established general trends throughout the department.
- *Diary completion.* A cross-section of employees was asked to complete a diary for a week, which looked at how they used their time and how often they were at their own desks or elsewhere.
- *Interviews.* These were held with a cross-section of the staff to follow up certain issues in more depth.
- *Room usage schedule.* These were focused on enclosed offices to see how often they were unoccupied.

Once completed, the post-occupancy evaluation allowed the consultants to identify rooms and areas that were underused. They were then able to make proposals for a revised departmental layout.

Comment
By using multiple methods, the consultants were able to establish which rooms/spaces were not being used to their full capacity, but more importantly they were able to question why this was happening. If they had used questionnaires or observational methods only, they may have drawn different conclusions. This could have resulted in a revised layout that did not meet the requirements of the department.

---

have been analysed try to prioritise the findings so that the most problematic or important areas are dealt with first.

3.4.5 *Presentation techniques*

Once data have been collected and analysed it will be necessary to present the findings in an appropriate format. The format selected will be dependent on how the information will be used. Some organisations will require the findings to be presented in a report that they can distribute to managers. Others will want a demonstration with overheads or flip charts. The facilities manager should really decide what method of presentation is suitable for the particular organisation. However, when presenting any findings, it is helpful to bear the following in mind:

- Try to prioritise the findings so that the most important items are dealt with first.
- Present the findings as simply as possible; overdetailed presentations will leave the reader/observer unsure of what the results are.
- Charts and tables help to get points across quickly and easily.
- Photos or videos can be useful to illustrate problem areas within existing buildings.

3.5   Conclusions

This chapter has focused on building evaluation from the users' perspective, primarily using post-occupancy evaluation techniques. This is a practical way of working to release the potential, identified in Chapter 2, of using an understanding of users as a springboard to facilities management excellence. Within this overarching context, the next chapter will consider alternatives and choices facilities managers should consider when organising their operations.

References

Becker, F. (1990) *The Total Workplace*. Van Nostrand Reinhold, New York. p. 263.

Ellis, P. (1987) Post-occupancy evaluation. *Facilities*, **5**(11), 12–14.

Kernohan, D., Gray, J., Daish, J. and Joiner, D. (1992) *User Participation in Building Design and Management*. Butterworth Architecture, Oxford.

Moleski, W.H. and Lang, J.T. (1986) Organisational goals and human needs in office planning. In *Behavioral Issues in Office Design* (ed. J.D. Wineman). Van Nostrand Reinhold, New York. p. 40.

Powell, J. (1991) Clients, designers and contractors: the harmony of able design teams. In *Practice Management: New Perspectives for the Construction Professional* (eds P. Barrett and A.R. Males). E. and F.N. Spon, London. pp. 137–148.

Preiser, W.F.E., Rabinowitz, H.Z. & White, T.E. (1988) *Post-occupancy Evaluation*. Van Nostrand Reinhold, New York.

Preiser, W., Vischer, J. & White, E. (1991) *Design Intervention: Toward a More Humane Architecture*. Van Nostrand Reinhold, New York.

Sanoff, H. (1968) *Techniques of Evaluation for Designers*. Design Research Laboratory, School of Design, North Carolina State University, Raleigh, NC.

Teicholz, E. (ed.) (2001) *Facility Design and Management Handbook*, McGraw-Hill, New York.

Tom, S. (2008) Managing energy and comfort: don't sacrifice comfort when managing energy. *ASHRAE Journal* (June), 18–26.

Zeisel, J. (1984) *Inquiry by Design*. Cambridge University Press, Cambridge.

Appendix: Post-occupancy evaluation data sheets

Date:	**POST-OCCUPANCY EVALUATION DATA SHEETS**	Ref:
	General information	

Building and department	
Room name/no.	
Purpose of room	
Brief description	
Is room size appropriate?	
Location within dept.	
Overall suitability	
Names of users	

	Sketch of room

User's additional comments	**Facilitator's comments**

Date:	POST-OCCUPANCY EVALUATION DATA SHEETS	Ref:

Wall finish	
Description	
Suitability	
Durability	
Maintenance	
Aesthetics	

Floor finish	
Description	
Suitability	
Durability	
Maintenance	
Aesthetics	

Ceiling finish	
Description	
Suitability	
Durability	
Maintenance	
Aesthetics	

Doors	
Description	
Suitability	
Durability	
Maintenance	
Aesthetics	

Date:	POST-OCCUPANCY EVALUATION DATA SHEETS	Ref:

Windows	
Description	
Suitability	
Durability	
Maintenance	
Aesthetics	

Lighting	
Description	
Suitability	
Durability	
Maintenance	
Aesthetics	

Power, communications and safety	
Name of item	Consider provision and location of following elements: electrical power outlets, data outlets, telephone points, fire alarms etc.

Furniture and equipment	
Name of item	What items of furniture/equipment are contained in the room? Are they well positioned? Does the room require further items of furniture/equipment?

II Organising for Facilities Management

4 Organising for Facilities Management

4.1 Introduction

4.1.1 Scope of the chapter

In this chapter we look more closely at organising FM. Organising can be thought of as the act of rearranging the various elements of an FM operation according to particular rules. Performance measurement is introduced at the outset as a key management tool. As such, it allows us to evaluate the merits of different FM arrangements. The chapter suggests that reliance on static measures of performance may thwart future growth. Consistent with the 'FM systems model' discussed in Chapter 1, a responsive measurement approach is put forward that is alive to changes in the external environment. Through a deeper understanding of relationship types, a contingent (i.e. one size does not fit all) approach is described that goes beyond cumbersome contracts and specifications. It recognises the increasing role of trust and shared values when organising service providers.

4.1.2 Summary of the different sections

- Section 4.1. Introduction.
- Section 4.2. The advent of performance measurement is explored as an organising tool. The section considers how our ability to measure has been pivotal in opening up the FM market, giving rise to increased competition and a panoply of insourcing and outsourcing options.
- Section 4.3. Considers the role of the 'intelligent client' in providing organising skills from within the client organisation. This then allows a full exploration of insourcing and outsourcing options.
- Section 4.4. Highlights the appeal and limitations of performance measurement in an increasingly complex FM environment. It suggests that overreliance on formal measurement can lead to an environment of mistrust and spiralling supervision costs. However, the advent of international performance measurement standards will help to alleviate inconsistencies in measurement approaches.

Facilities Management: The Dynamics of Excellence, Third Edition. Peter Barrett and Edward Finch.
© 2014 John Wiley & Sons, Ltd. Published 2014 by John Wiley & Sons, Ltd.

- Section 4.5. Explores how varied 'relationship approaches' can be used to meet the FM challenge. Rather than restricting the debate to outsourcing versus in-house, this section considers a number of ways that FM can be organised based on relationship approaches, whether these be internal or external.
- Section 4.6. The merits of the service procurement strategy is frequently based on cost efficiency and value-adding considerations. However, the dynamic nature of FM means that static tie-ins to a particular contractor or mode of operation can be costly. Flexibility in its various forms is examined as the key yardstick for avoiding such inertia.
- Section 4.7. This section considers the dual character of facilities management as both a product (hard FM) and service (soft FM). It suggests that an organising approach appropriate to hard FM may not be applicable in soft FM, where the role of the customer in the coproduction process is key.
- Section 4.8. Examines how the advent of the 'experience economy' makes the challenge of organising FM ever greater, as organisations continuously seek reinvention through flexible environments.
- Section 4.9. Conclusions.

4.2 Performance measurement

The management of buildings has historically been an activity undertaken by owners or occupiers themselves. Long-term activities such as relocation or new-build activities were performed or overseen by people within the organisation. Likewise, day-to-day operations such as maintenance, space planning and other service provisions were undertaken in-house. Sometimes these tasks were done well, at other times they were done badly. All such activities were undertaken alongside other company activities, irrespective of how relevant such activities were to the business of the organisation, whether it be producing products or delivering a service. The absence of any mechanism for evaluating how well an outside organisation might perform such facilities operations (as well as the scarcity of providers) meant that such processes were almost universally retained in-house.

The latter part of the twentieth century saw the introduction of key management techniques such as total quality management and business process re-engineering. More recently, the application of lean principles in facilities management has brought a spotlight on business activities that had previously been the source of substantial waste. The shortcomings of existing practices were brought into sharp relief – the days of the untimely repair or the casual property decision were over. Having 'commoditised' activities such as facilities management, based on modern management techniques, the possibility arose of determining how well an outside provider might provide such a service. Indeed, the emergence of an increasingly sophisticated performance measurement approach allowed outside providers to compete in a much more transparent and sizeable market. This process, known as outsourcing, has come to define the subject of facilities management (rightly or not). What is undeniable is that the development of measurement tools for evaluating cost, risk, value and flexibility were essential prerequisites to the modern-day emergence of facilities management, whether in-house or outsourced.

4.3   Organising the FM team

In the face of growing budgetary constraints, many modern organisations are choosing to adopt a 'thin client' organisational structure in relation to facilities management, retaining only a minimal in-house management capability. This is in response to organisational demands and the need to focus on core business.

However, is it possible for an organisation to relinquish all responsibilities using this approach? The dangers of 'throwing the baby out with the bathwater' are well documented in relation to outsourcing, whereby clients can be held to ransom without the necessary in-house skills to evaluate and monitor service providers.

Are there a set of organising capabilities that need to be sustained within an organisation, irrespective of whether an outsourcing or in-house operation is applied? Williams (1996) suggests that there is a core capability that needs to be retained in-house. This facilities management capability can be described as the 'intelligent client'. Essentially, this provides a senior management capability that is able to represent the interests of the organisation. In order to do this, three facets need to be addressed:

1.   Sponsorship (policy and strategy) involving:
     (a)   Creation (support from within and outside of the organisation)
     (b)   Strategy formulation
     (c)   Changing
     (d)   Directing
2.   Intelligence (understanding and monitoring) addressing:
     (a)   Customer objectives
     (b)   Customer needs
     (c)   Technology
     (d)   Service delivery
3.   Service management (contract management) including:
     (a)   Agency
     (b)   Task management
     (c)   Contract management

Williams (1996) goes on to argue that 'the proper allocation of resources to each of these facets is absolutely critical to the achievement of cost effective facilities management'. He suggests that many organisations often possess only one or two of these three essential facets (sponsorship, intelligence and service management).

Organisations invariably focus on the procurement process (service management role) as an opportunity to 'squeeze out' savings. However, the opportunities for budgetary control (enabled through the forward-looking sponsorship role) and the opportunities for value engineering (enabled through the intelligence role and an understanding of need) are two equally important areas. Both of these facets do not need to come at the expense of the outsourced service provider's profit margin. A mature partnering relationship with an outsourced provider should explore the possibilities for budgetary control and value engineering.

The generic 'systems' model discussed in Chapter 1 clearly aligns with that of Williams (1996). The Sponsorship role describes the 'outside and future' perspective addressing changes to the external environment. The Intelligence role, in contrast, is more concerned with scanning developments in the core business, although again it is future oriented. Finally, the Service Management role focuses much more on the 'inside and now' supported in turn by IT and benchmarking.

With the increasing reliance on the 'intelligent client', the requirement for measurement becomes pivotal to the success of the client organisation. Without a measurement tool that truly reflects the strategic long-term impact of the FM team's performance, it is impractical for the client organisation to establish a viable system. The following sections consider in more detail how such a measurement strategy might be used to guide the organisation, both as a singular unit and as an enterprise.

4.4   Facilities management and measurement

The diversity of procurement routes and the increasingly competitive market in facilities management can be attributed to the growing confidence in performance measurement. Innovations in FM standards and the development of corporate-wide IT-based FM systems have accelerated this process. Whilst the debate over in-house or outsourced provision remains largely unresolved, performance measurement systems have enabled both approaches to prove their worth. The old adage 'you cannot manage what you cannot measure' was expanded upon by Halachmi (2005: p. 503):

1. if you cannot measure it you do not understand it;
2. if you cannot understand it you cannot control it;
3. if you cannot control it you cannot improve it;
4. if they know you intend to measure it, they will get it done;
5. if you do not measure results, you cannot tell success from failure;
6. if you cannot see success, you cannot reward it;
7. if you cannot reward success, you are probably rewarding failure;
8. if you will not recognize success, you may not be able to sustain it;
9. if you cannot see success/failure, you cannot learn from it;
10. if you cannot recognize failure, you will repeat old mistakes and keep wasting resources.

Of course, performance measurement itself throws up a number of issues and can be the cause of resentment in organizations. Mistrust is a symptom of a system reliant on measurement to get things done. The fourth item identified by Halachmi (2005: p. 503) previously, clearly fits within the McGregor (1960) Theory X model of management styles, which assumes that employees lack initiative and will avoid work if they can, particularly if it involves the expenditure of resources. The corollary of this is that management consider that employees need to be closely supervised and comprehensive systems of control developed. A similar assumption prevails regarding the disposition of external contractors. It assumes that, only through close monitoring (looking over the shoulder), is it possible to ensure performance improvements.

Taking the alternative Theory Y view of employees and external contractors, performance measurement acquires an entirely different role. Theory Y assumes employees are naturally ambitious and *self-motivated* and possess *self-control*. It is assumed that employees (as well as contracted staff) enjoy their work responsibilities such that work is as natural as play.

Only by possessing a clear position regarding Theory X or Theory Y thinking is it possible to identify a coherent performance measurement approach. Theory Y provides a positive set of beliefs about FM employees and contractors and the possibilities that this creates. Theory Y performance measurement systems are more likely than Theory X systems to develop a climate of trust necessary for human resource development and partnering. Inevitably, the extent to which Theory X or Theory Y prevails is not just dependent on the management style imposed by the client. It may reflect a strongly embedded culture within the service provider organisation. Furthermore, for outsourced contracts, the type of outsourcing relationship that exists also has a bearing on the dominance of Theory X or Theory Y thinking. Specifying a contract that encourages one or other behaviours can make the emergence of either theory a self-fulfilling prophecy. Inevitably, trust has become a progressively more important part of complex contracts involving intangible assets. As we shall see in this chapter, the importance of intangible aspects of facilities management makes performance measurement ever more challenging.

4.4.1 Commoditisation through standards

In the seminal work of Davenport (2005) he proclaimed that 'business processes, from making a mousetrap to hiring a CEO – are being analysed, standardized and quality checked. That work, as it progresses, will lead to commoditization and outsourcing on a massive scale'. The facilities management industry is at the forefront of this change. The promise of cost reductions, a more tightly controlled balance sheet, greater flexibility as well as the lure of access to a more diverse and specialized labour market has proved compelling for many organisations. Commoditisation in FM through the introduction of national and global standards has arguably made this ever more possible. Clients can now supposedly 'shop with confidence' based on the existence of a common set of processes (e.g. BS EN 15221-5:2011, Facility management. Terms and definitions). However, whilst commoditisation bears opportunities for owners of facilities through reduced switching costs, for those involved in providing a facilities service, commoditisation might be seen as a 'double-edged sword'. Commoditisation, in a less positive light, describes a process whereby goods or services that have a definable economic value, and which previously were distinguishable in terms of features (brand offering), end up becoming simple commodities as seen by the market. The FM market might indeed be undergoing a significant move from one being monopolistic (i.e. only one possible provider – the in-house team) to one of perfect competition – from a differentiated local market to an undifferentiated global one, competing solely on price. Not all observers identify with the upbeat view of Davenport (2005) regarding the assumed benefits of outsourcing. In an earlier study by Cant and Jeynes (1998) involving over 20 change management/outsourcing initiatives (over one-third being in-house performance improvement initiatives), the authors arrived at the question:

> Is it possible that market testing is a too-hasty solution to the undeniable need for rapid change and improvement and that, with slightly less urgency, departments would have been better served through the application of innovation rather than market testing?

For the UK public sector, Cant and Jeynes (1998) advocate changing the process model to allow for improvements in existing services prior to a costly market-testing exercise (outsourcing). To this end, we can see that performance measurement and the formulation of standards has an equally vital role to play.

Inevitably, the landscape of facilities management will be transformed, driven by the development of new process standards. Businesses will find it increasingly possible to assess whether an FM capability can be improved by outsourcing or an internal change process. It is suggested that the cost of service delivery will fall in the face of greater competition and FM, through commoditisation, will be able to reach across boundaries (for small FM organisations this will provide reach across a wider regional area, whilst for larger FM operations it will enable truly international reach). Three fundamentally different types of standard support the commoditisation process in FM as shown in Figure 4.1:

1. The standardisation of processes (what do you do?)
2. Process performance standards (how well do you do it?)
3. Process management standards (how do you do it?)

1. Standardisation of process

Significant efforts have occurred over the last decade in attempts to define the precise nature of FM processes. This extends to our common shared understanding of work packages within the industry, as well as the set of activities required to fulfil a particular task.

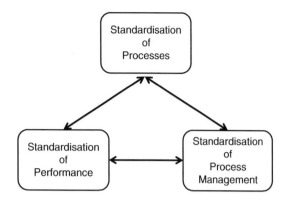

Figure 4.1   Standards and commodification in facilities management.

Most influential in the European context is the European Committee for Standardisation (prEN 15221). This committee has put forward a number of standards in relation to process, which in turn have been embraced in country-specific standards. In the British Standard these publications include:

- BS EN 15221-1:2006: Facility management. Terms and definitions
- BS 8536:2010: Facility management briefing. Code of practice
- BS EN 15221-4:2011: Facility management. Taxonomy, classification and structures in facility management
- BS EN 15221-5:2011: Facility management. Guidance on facility management processes.

Standardisation of process begins with establishing a 'shared meaning'. In the case of facilities management standards, taxonomy and the definition of terms and measurements provide a key starting point to process standardisation. In so doing, 'a set of classified facility products will, if widely adopted, allow organisations to align internal structures/costs and, over time, benchmark with other similar companies with increased certainty'.

Given the complexity of facilities management in terms of regional variation and asset-specificity (i.e. every building is different), common standards continue to represent a challenge. Moreover, activities such as workplace planning can cut across numerous cross-functional teams including IT and human resources. Despite these challenges, standards offer the possibility of unambiguously defining process, such that variability in how organisations define an activity can be reduced, the result being that it is increasingly possible to market test a facilities management activity. As well as enabling the communication of how FM operates, it enables the smooth hand-off across process boundaries. Process standardisation is an essential prelude to benchmarking (performance measurement).

2. Standardisation of performance measures (benchmarking)

Having identified processes, the challenge in FM remains as to how such processes and configurations might be evaluated and compared? In other words, how well are we doing? BS EN 15221-7, Facility management Part 7: Guidelines for performance benchmarking, provides the draft European standard for benchmarking in FM. Benchmarking can provide organisations with an understanding of the scope for improvements in an FM operation based on a systematic basis of comparison. The draft standard suggests that, despite being a commonly used tool in the industry, it 'has often been misused and misunderstood within facility management' (BS EN 15221-7, p. 4). The standard identifies the wide-ranging purposes to which FM benchmarking can be applied. These include:

- Identification of improvement options
- Resource allocation decisions
- Verification of legal compliance
- Assessment of property performance
- Evaluation of floor space usage
- Budget review and planning

Benchmarking approaches vary in terms of reach (intraorganisational, inter-organisational, regional, national and international) and periodicity (one-off, periodic or continuous) and fall into one of several categories described below (each of these categories can be identified within the 'generic FM systems' model illustrated in Chapter 1 in relation to emphasis on 'outside and future' or 'insider and now'):

Strategic benchmarking is driven by the need to improve overall performance. Such benchmarking involves high-level aspects of FM delivery such as new service development, change management impacts and core competencies.

Performance or competitive benchmarking involves the analysis of performance in relation to key products (e.g. property in relation to occupancy levels in the hotel sector) and services (e.g. helpdesk operation). Energy and sustainability benchmarking are perhaps the most widespread form of this benchmarking approach in FM (see LEED, BREEAM, and Energy Star).

Process benchmarking involves a focus on specific FM processes. It allows the identification of gaps in performance compared to others in the FM sector and frequently gives rise to short-term benefits.

Functional benchmarking is perhaps the most unexplored form of benchmarking in which partners are sought from different FM sectors to allow analysis and comparison. Whilst their business type might be different, common challenges often arise from similar tasks or the requirement to deliver similar services. One example of such a functional analysis is post-occupancy evaluation, wherein the functional performance of a school facility might be compared with that of exemplar hotel facilities or hospitals that are known to excel in one or more respect.

Internal benchmarking involves benchmarking FM operations from within the same organisation. Whilst it benefits from ready access to comparable data in the same organisation, it does not provide the same kind of insights of best practice possible through external benchmarking.

External benchmarking requires performance measurement of other organisations whose FM practice has been identified as best in class. To obtain the greatest benefit it is recommended that comparators are chosen from a wide pool of industries, thus capturing new approaches.

International benchmarking. In situations where 'best in class' is rare, it may be necessary to examine a larger pool of global FM practice. The limitations of such an approach may be the lack of applicability of 'lessons' from cultural and national FM settings significantly different from the one in question.

All benchmarking techniques rely on the existence of a *standard* metric (indicator) to measure performance (cost per square metre, productivity per unit of measure, cycle time of x per unit of measure or defects per unit of measure). This then provides the basis of comparison.

Also of relevance to measurement is the standard related to area and space as outlined in the British Standard:

- BS EN 15221-6:2011: Facility management. Area and space measurement in facility management.

3. Standardisation of process management

The last of the three aspects of standardization enabling the commoditisation of facilities management addresses the question of quality and quality control. Several European standards have recently emerged that seek to establish 'how well do you do it?' Building on the influential principles endorsed in the ISO 9000 standards, which are relevant to any business or organisational activity, several that are specific to the facilities management domain have emerged. Examples related to the UK context include:

- British Standard BS EN 15221-3:2011 (Facility management. Guidance on quality in facility management).
- British Standard BS 8572:2011 (Procurement of facility-related services) provides guidance on how to procure facilities services in a consistent and thorough manner. As such, it addresses considerations related to value for money and end-user satisfaction.
- Publicly Available Specification (PAS) 55, which provides guidance on asset management wherein asset management is defined as the 'systematic and coordinated activities and practices through which an organization optimally and sustainably manages its assets and asset systems, their associated performance, risks and expenditures over their life cycles for the purpose of achieving its organizational strategic plan'. The specification is relevant to a broad range

of infrastructure assets beyond conventional facilities management (e.g. process facilities, industrial and civil engineering activities). A much needed characteristic of this specification is that it attempts to go beyond a simple analysis of whole-life cost trade-offs based on fixed assumptions about the asset's life. It considers issues related to risk exposure, performance, maintenance and obsolescence that vitally affect the economic performance of the asset.

What is evident from the significant efforts in standards is that facilities management is becoming a mainstream industry capable of exploiting the market accessibility that comes about from commoditisation. In the next section we consider some of the potential pitfalls of commoditisation and the checks and balances necessary to avoid some of the negative effects that can arise.

The area of information exchange between construction industry professionals has also been the subject of process management integration. Indeed, the advent of 'building information modelling' discussed in Chapter 7 relies on such process management standards. A useful guide for this standardisation is given in the publication:

- BS 1192:2007: Collaborative production of architectural, engineering and construction information. Code of practice.

4.5 Relationship approaches to FM

Whether the 'intelligent client' is considering in-house procurement or outsourcing, success pivotally depends upon the relationship that is established with the service provider. Using the parlance of Agency Costs Theory, this relationship involves a client (principal) delegating the performance of services to an external vendor (agent). The consequence of this is that costs arise that in turn need to be managed using a variety of control mechanisms. The nature of this relationship can take several forms.

Some researchers on the subject of outsourcing have chosen to focus on 'relationship types' rather than the polarised discussion relating to outsourcing versus in-house operation. In so doing, they highlight strategies that fit with the characteristics of the client organisation and the service concerned. In the realm of facilities management, Hui and Tsang (2004) examined insourcing, out-tasking, outsourcing for cost saving and outsourcing for capability, seeking to provide guidance on each approach. In the more general area of IT outsourcing, Kishore et al. (2003) developed a relationship model (Four Outsourcing Relationship Types) based on two parameters (level of substitution and level of strategic importance). Substitution in this context refers to the extent to which the service provider (vendor) has decision-making authority over the operations; the involvement of the vendor in the planning and implementation stages; the extent of shared ownership relating to hardware and software; and the extent to which personnel from the vendor organisation replace those from the client organisation.

Using the two parameters (substitution and strategic importance) they were able to differentiate four relationship types (Support, Alignment, Reliance, Alliance). Sia, Koh and Tan (2008), in contrast to the work of Kishore, chose to look at the specific issue of 'flexibility' in-service procurement. Based on various economic theories they were able to recommend 'manoeuvres' that would help with managing flexibility in the outsourcing process. These manoeuvres included pre-emptive, protective, exploitative and corrective manoeuvres.

In relation to FM, we can apply the principles of 'systems theory' outlined in the generic FM model of Chapter 1, using two key parameters that dictate what form FM relationships typically take. These two parameters are:

- focus on the internal/external environment;
- focus on the present/future.

Using these two parameters it is possible to create a relationship matrix involving four distinct types of FM relationship (between vendor and client). These four types of relationship (the four C's of FM) are identified in Figure 4.2. They include *commoditisation*, *customisation*, *consultation* and *collaboration*.

In the bottom left-hand corner of the matrix is *commoditisation*. It represents the dominant model of FM procurement in the industry, with the emphasis on increased efficiency and cost savings. This in turn is driven by a measurement model that is internally focused, seeking to attenuate (reduce) variety from the external environment. Similarly, the *commoditisation* approach is focused on the 'here and now' rather than any long-term perspective. Management control in the *commoditisation* model relies centrally on the 'water-tightness' of the contract. The *customisation* model (which is shown in the top left-hand quadrant of the matrix in Figure 4.2) recognises the emergent and changing demands of an organisation. Rather than relying on the fixity of the contract, other approaches are

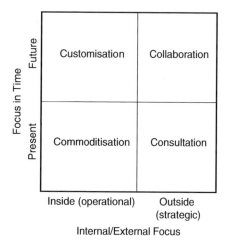

Figure 4.2   The four C's of FM relationship types.

used to achieve management control. This is largely because of the inflexibility imposed by contracts for a customised environment where change management becomes an inevitable and ongoing concern. In the *customisation* model the focus turns to the future changes from within rather than the present 'fixed' scenario.

Moving to the right-hand side of the matrix (*consultation* and *collaboration*), sensitisation to the external environment becomes the dominant theme. In the case of *consultation*, external experts (to include in-house expertise that is externally oriented) provide an outward-facing orientation that reflects a leading edge understanding of their own field. Such consultants are typically involved in discrete projects, with specific time horizons. Their contribution, like the *commoditisation* approach, is largely focused on the 'here and now', ensuring that the organisation is 'up to speed' with current best practice. Typically, the output from a *consultation* relationship takes the form of a prescriptive solution that suits the specific demands of a particular project. Another form of relationship type is the collaborative relationship that seeks to increase (amplify) the variety of an FM organisation by being externally oriented. Unlike the *consultation* approach, the collaborative approach relies on an ongoing relationship with a service provider (that has high levels of awareness regarding external practice and leading-edge know-how). Whilst the *consultation* relationship type is concerned with planning and getting it right first time, the collaborative model relies upon an ongoing and adaptive approach. This is based fundamentally on shared goals and high levels of trust. Whilst the contract provides a necessary prerequisite, in the collaborative model it does not represent the basis of the relationship. Instead, an output-based approach rather than a prescriptive approach prevails.

No one relationship is ideal for all organisations or service requirements. Only by considering the individual characteristics of each relationship type is it possible to assess their appropriateness. With this in mind, the attributes of each relationship type (the four C's) are considered in turn.

4.5.1 Commoditisation

This describes a relationship in which the service provided is seen by the FM market (and the client) as a tradeable commodity. In this context, the market for the service is considered to be largely undifferentiated and it is possible to achieve something close to perfect competition. The existence of FM standards ensures that comparisons can be made between providers, with the preferred provider being chosen on the basis of conformity and cost.

With the emphasis on price competition, *commoditisation* relationships are characterised by:

- Reliance on the **contract**, which serves as the main control mechanism. The robustness of such a contract depends upon the ability to assimilate changes during the course of contract.
- A **prescriptive** approach driven by the 'intelligent client' that allows direct comparison of potential service providers in an undifferentiated context

(level playing field). A prescriptive specification involves contracts that not only outline the performance required but also may prescribe the design, products and workmanship required to deliver that performance. One criticism of this approach is that it may lead to suboptimal solutions that do not harness ideas and expertise possessed by the service provider themselves. However, the ability to commoditise a service enables the client to take full advantage of an FM market.

- A **universal** approach that can equally be applied to in-house operations.
- Emphasis on **efficiency**, with the successful provider being able to do the same thing as others but at a lower cost.
- Relatively **low substitution** levels in terms of displacement of in-house expertise, equipment and resources, thus allowing easy switching between service providers.
- Relatively **low setup costs**.
- Success depends upon the existence of a **mature process** with established practices and workflow patterns.
- Performance measurement based on process/task measurement and monitoring.

The *commoditisation* approach is widely seen as most effective in services that are not subject to rapid rates of change or high levels of complexity. The theory of Transaction Cost Economics (Williamson, 1985) provides a useful lens for understanding costs as seen from the *commoditisation* perspective. In a perfect market the single factor determining the choice of FM organisation is the transaction cost. As seen from the perspective of transaction cost economics (TCE), services involving high asset specificity, high transaction frequency and uncertainty are considered most appropriate for in-house delivery. However, in the FM market, the widespread evidence of commodification in outsourcing suggests that high transaction frequency does not deter clients from considering outsourcing as an option.

Case Study 1: Commoditisation relationship

A client organisation is responsible for managing several hundred high street banks nationally and recently awarded a maintenance management contract. Being an experienced client, it already has mature processes for prescribing maintenance management activities. Furthermore, it has adopted widely used standards in relation to FM processes. As a result, it was able to tender to a large market. The successful service provider was chosen primarily on price. Price transparency on the part of the vendor was achieved by conformance to benchmarking standards. Furthermore, interoperability in IT systems provided a key opportunity for minimising start-up costs and ensuring effective reporting. Switching costs and supplier dependency were of particular concern to the client in ensuring future flexibility.

4.5.2   *Customisation*

Every client organisation is different: *customisation* seeks to accommodate these differences by providing a tailored FM solution. Invariably, *customisation* occurs over time as organisations attempt to align the FM provision with the process requirements, cultural demands and policies of an organisation. Moreover, many organisations wish to appear distinct from the competition, seeking to avoid a 'vanilla' solution. The way in which FM is delivered may form part of this picture. Facilities managers often encounter increased costs associated with facility customisation. A common antagonism with signature architects is that of bespoke buildings, which may leave the facilities manager with increased repair and replacement costs owing to untried/ untested design solutions and materials that are not 'off-the-shelf'. However, the fluidity of customisation enables future requirements to be assimilated. Added to this is the building stock itself, which is often far from being a blank canvas, encompassing varied ages and construction type. This asset specificity often calls for a customised approach. Some of the features of a customised solution include:

- Emphasis on **effectiveness**.
- Performance measurement is **outcome based**.
- The contract is not the sole means of monitoring and control.
- **Incentivisation** using various mechanisms ensures alignment of objectives.
- High levels of **substitution** in terms of expertise, equipment and infrastructure on the part of the service provider.
- Higher levels of **trust**.
- Longer-term contracts compared to *commoditisation* relationships.

Asset specificity often demands a customised approach rather than a commoditised approach. In the case of FM, this is particularly true where property portfolios are often heterogeneous and unique to the organisation. However, Sia *et al.* (2008) point out that asset specificity is often incorrectly perceived as being an uncontrollable factor. Specifically, they suggest that the 'client organizations can consciously choose to minimize customization, and instead "live with" the generic or standard product offering of vendors' (Sia *et al.*, 2008: p. 411). Agency Costs Theory (Jensen and Meckling, 1976) substantiates the *customisation* approach, which focuses on the principal–agent relationship. In order to make such a relationship work, certain costs have to be foregone. The theory proposes various control mechanisms such as variable pricing, performance-based pricing and profit sharing, which serve to incentivise the contractor (by aligning interests) and avoid problems such as adverse contractor selection and underperformance.

> **Case Study 2: Customisation relationship**
>
> A hotel chain seeks to provide security services for hotels dispersed nation-
> ally. It sees FM provision as being close to core business and a mechanism
> for differentiation in the market. It has sought to appoint a new service
> provider that understands the subtle relationship between security and
> reception services. Hotel branding is a central component of the client's
> strategy. At this point in time the client is undecided as to whether to
> outsource or to use an improved in-house solution. The appeal of an
> in-house solution is that existing staff already know the characteristics of
> the hotel estates from a security perspective. However, further work is
> required to improve 'front of house' customer interaction. The dilemma is
> whether an outsourced provider can fully adopt the customised approach to
> reception/security services within the hotel chain. Set against this are con-
> cerns as to whether existing security staff are able to assimilate new working
> practices required to ensure good customer engagement.

4.5.3 Consultation

The *consultation* relationship approach is located in the bottom right-hand
quadrant of the matrix in Figure 4.2. Its position reflects an orientation towards
the external environment rather than preoccupation with internal operations.
Specifically, it seeks to increase internal responsiveness to a changing external
world by means of consultant expertise. The characteristics of a *consultation* relationship are as follows:

- Generally, project-based.
- Concerned with securing access to world-class technical expertise.
- Performance is usually based on behaviour-based control mechanisms as it is
  difficult to measure discrete outputs.
- Relies upon access to in-house expertise in order to document existing
  systems and requirements prior to the development of any new system.
- Generally the level of substitution is low and the relationship does not involve
  a transfer of assets, equipment or manpower.
- Being project based, the focus is on the 'here and now' and the delivery of
  specific projects.

Underpinning the *consultation* relationship model is a theoretical perspective
known as Social Exchange Theory, which assumes that interorganisational
exchanges are bound up in social relationships (Blau, 1964). In other words, the
existence of a strong social relationship between the client and vendor ensures willingness of the parties to adjust to changing demands (Poppo and Zenger, 2002).

Unlike the *commoditisation* relationship, the *consultation* model (involving 'thinkers' rather than 'doers') leads to intangible and often ambiguous outputs. Control and management in this context is usually based on behaviour-based mechanisms. This involves the ongoing reinforcement of promises and expectations by means of social processes founded on reciprocity and cooperation.

Case Study 3: Consultation relationship

A charitable organisation possesses an array of historic buildings located in a renowned lakeside area. These heritage buildings are subject to various controls by national bodies and local authorities. Such controls are designed to ensure that the historic integrity of the buildings is maintained. The client must ensure that the buildings are viable as business propositions as well as meeting conservation demands. As part of a broad-ranging effort to make the property portfolio income generating, a major series of refurbishments are proposed, with energy conservation at its heart. The recent appointment of a consultancy with varied expertise in architectural conservation, energy efficiency, sustainability and workplace design is seen as a major departure for the organisation. Previously, property assets were not seen as being capable of generating income. The infusion of new ideas from the consultant team is seen as a way of harnessing latent potential within the estate.

4.5.4 Collaboration

Often thought of as being the most mature type of relationship encountered in FM, *collaboration* involves a close long-term working relationship with an FM provider. Potentially such a relationship can have a significant impact on the future of both the client and provider. A collaborative relationship is characterised by:

- An approach with **viability** in mind, seeking to ensure the long-term survival of both parties.
- High levels of **substitution**, with much of the resources and manpower provided by the vendor.
- **Common objectives** and missions designed to ensure goal symmetry.
- Use of a **common communication system** (e.g. Building Management system, Building Information Management (BIM) system or Building Maintenance Management system).
- Outcomes that are often difficult to specify. As a result, performance and monitoring are typically longer term and reliant on trust.
- Low levels of contractual control are exercised.
- High **switching costs**.

The expertise of the vendor 'partner', often operating with multiple clients, provides considerable reach regarding the external environment. This provides access to best practice within the relationship. Furthermore, the long-term nature of the relationship legitimises investment in interoperability and common systems. Unlike the motive for interoperability in the *commoditisation* relationship (allowing ready switching of suppliers), interoperability in a collaborative relationship is seen as a longer-term commitment in the pursuit of a resilient FM system.

The theory of Entrepreneurial Actions originally formulated by Schumpeter (1934) is strongly aligned with the collaborative relationship model. It argues that entrepreneurs are the main agents of change, introducing new methods of production and new products as well as other forms of innovation that give rise to increased economic activity. Entrepreneurship is described by Schumpeter (1934) as the process of 'creative destruction' such that the entrepreneur (in this case, the vendor in partnership with the client) continually displaces or destroys existing methods, replacing them with new ones. It is this future-oriented, proactive and outward-looking approach that characterises the *consultation* relationship, positioning it in the upper right-hand quadrant of the matrix in Figure 4.2.

Case Study 4: Collaboration relationship

Two organisations (client and vendor) have come together in a long-term partnering arrangement. The supplier in the relationship is in a position to invest capital to allow the incorporation of state-of-the-art building management systems (BMSs) throughout the property portfolio. An arrangement of gain sharing has been made for energy savings accruing over the next 10 years. It has also been seen as an opportunity for the two partnering organisations to establish a new IT system based on emerging interoperable technologies. Given the uncertain climate that the client (a pharmaceutical organisation) is likely to encounter in the foreseeable future, engagement with a provider that is future oriented is of key importance. High rates of change in the client organisation mean that a flexible approach to service provision is necessary, with a collaborative relationship based on trust rather than adherence to a contract.

4.6   Flexibility and the FM organisation

One of the burning questions facing the intelligent client is 'what procurement route provides the greatest flexibility for an uncertain future?' Sia *et al.* (2008) attempted to understand outsourcing flexibility, recognising that there are a number of dimensions to this concept. Their study was based on 171 projects in Singapore and gave rise to four identifiable dimensions of outsourcing flexibility:

- Robustness, which refers to the variability of service capacity (e.g. fluctuations in occupancy levels, number of faults reported, urgent requests).

- Modifiability, which describes the alteration to service characteristics in the face of changing facility/organisational requirements (e.g. change of use, changes in operating hours).
- New capability, which involves the incorporation of novel or innovative capabilities arising from radical shifts in the building occupiers' (clients') business requirements (e.g. meeting new building energy performance standards, enabling new ways of working, satisfying requirements of new health and safety legislation).
- Ease of exit, which describes the ability to switch to another provider or to return to an in-house service.

Each type of flexibility presents different challenges. Thus, whilst modifiability and robustness can be addressed within the context of an existing service, new capabilities necessitate access to new skills and technology. These new sources of expertise typically arise in relationships that are attuned to external changes. Considering the four C's of relationship types encountered in FM, the *collaboration* and the *consultation* relationship types provide environments for incorporating new capabilities (i.e. outward facing).

As a result of outsourcing, vendors can exercise a high level of control in FM provision. Resource Dependency Theory (RDT) (Pfeffer and Salancik, 2003) examines how this unwelcome overdependence can arise and how it can be mitigated. Multiple sourcing is one way of addressing such risks. However, another way of actively managing such risks is to ensure vendor interoperability. Having the option of changing the provider at the end of the contract is often of paramount concern. Ease of exit is one of the main reasons for adopting a *commoditisation* relationship. In principle, the adoption of a mature process and reliance on recognised standards enables the client to switch to another vendor without significant costs.

4.7 Tangible and intangible FM

What are FM relationships trying to achieve? The unprecedented growth of facilities management in the 1980s and 1990s saw the switch of many real estate organisations into facilities management, recognising the increased business opportunity, which contrasted markedly with the lacklustre performance of the real estate business at the time. For many such organisations this was a hazardous (and frequently a failed) venture into an area fundamentally different from the 'bricks and mortar' concerns of real estate investors. To understand this duality of 'product' and 'service', which is central to the FM ethos, is to understand true performance measurement.

One commentator, observing the state of the healthcare industry in the UK, asserted that 'we know everything about treatment and nothing about care' (Anon). This effect often arises from the challenge of measuring those factors that are often difficult to measure – the intangible elements. In the words of Albert Einstein: 'Everything that can be counted does not necessarily count; everything that counts cannot necessarily be counted.' Such accusations could also be levelled at many facilities management initiatives, driven by recent pressures to reduce costs and implement process-driven solutions.

Is there a difference between an organisation that produces something that can be seen, touched and held, and managing an organisation that produces something that is perceived, sensed and experienced? That is the key question as we move from an FM industry preoccupied with the physical asset to one addressing the unique challenges of service provision. We can think of a service as an act carried out by a service provider that may be associated with a physical product such as a building although the action is essentially intangible and does not result in something that can be owned.

Several key differences between managing a service and managing a product have been identified by Gronroos (2000), Pine and Gilmore (1998) and Zeithaml and Bitner (2000):

- **Intangibility**. Services are essentially intangible and thus cannot be 'inventorised' or stored. This 'perishable' characteristic means that you cannot draw on replacement parts or provisions in response to an unanticipated fluctuation in demand.
- **Variability**. Services exhibit variability and are thus difficult to control because of unanticipated customer demand.
- **Heterogeneity**. The service interaction itself may differ from one instance to another. Both the customer and provider may vary in terms of training, intelligibility, style, empathy and time pressures.
- **Simultaneity**. Unlike a production activity (e.g. building maintenance), pure service interactions necessarily involve the customer at the point of production. As a consequence, the customer witnesses shortcomings in the service at first hand.

Whilst not all facilities management activities occur in the presence of building occupiers, we increasingly see hard FM impacting on the experience of customers. Retrofits and refurbishments in high-change organisations are often undertaken in close proximity to the building users, who may themselves have had to relocate temporarily to accommodate the change. Thus, activities that traditionally were seen as 'back-office' activities are impacting on the user organisation, often in unintended ways. The practice of introducing daytime office cleaning is a further example of how facilities management is becoming an increasingly front-of-house activity, making demands on customer service skills.

User participation is perhaps the extreme end of the product–service spectrum, such that building users are empowered to decide on and manage different aspects of their environment. Whilst this trend has been welcomed in areas such as thermal comfort, choice of furniture systems and flexible partitioning, other areas such as hot-desking and office cleaning may be less appealing from the user perspective. In hard FM, the decision-making process tends to be long term, intractable and less amenable to user participation. Even when such users are involved in such design decisions, they may only be serving a 'proxy' role for the eventual users because of the time taken for the delivery of the product (i.e. building).

FM services are essentially perishable and have a 'sell-by' date. If it is not used, the service offering perishes. An example would be staffing of a facility helpdesk operation. Reason would suggest that staffing of such an operation needs to accommodate sudden rises in demand because of unexpected events or random variations. A consequence of this is that there may be significant idle times. However, given the perishable nature of services, a lost service offering cannot be recovered. This is in direct contrast to products that remain in perpetuity and are available for use as required. In practice, most facilities management products/services often have both a tangible and intangible element somewhere along a continuum.

4.7.1 Hierarchy of needs

If we are to measure performance effectively and tackle the espoused intention of supporting the needs of an organisation, we need first of all to be able to identify such needs and subsequently be able to identify the extent to which such needs are met through facilities management. Pivotal to our understanding of organisational needs is the early work of Maslow. In his now celebrated work on *Motivation and Personality* (Maslow, Frager and Fadiman, 1970) he examined the main drivers behind individual human behaviour. He proposed a 'hierarchy of needs' such that basic human needs must be satisfied before the pursuit of higher-order needs associated with personal fulfilment.

Whilst the original theory has been the subject of criticism by some modern-day management theorists, the principles have largely withstood the test of time, albeit in modified forms. The five levels of the hierarchy represent an evolutionary path in human development, as shown in Figure 4.3.

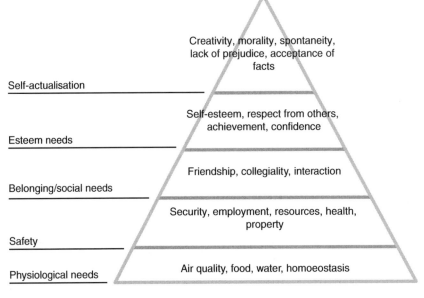

Figure 4.3   Maslow's hierarchy of needs (Maslow, 1943).

The central tenet of this model is that individuals will only pursue needs at a given level once all of the needs of the preceding level have been satisfied. Further, if the more basic needs cease to be fulfilled, the attention of the individual reverts to efforts to satisfy these more basic needs.

- *Physiological needs.* The hierarchy begins by addressing physiological needs at the base. These constitute basic human needs. In the built environment, the capability of delivering homogeneous comfort conditions, adequate lighting and building integrity are features that will determine the satisfaction of these needs, as shown in Figure 4.3. The inability to satisfy these basic needs will result in an unusable space.
- *Safety needs.* Having satisfied the rudimentary requirements of basic human needs, individuals will seek to satisfy safety needs. This term encompasses a broad range of phenomenon that covers both physical and psychological well-being. Resource requirements including access to equipment (such as printing) and access to other people apply to safety needs (e.g. through appropriate space planning). Access to natural light and the presence of acceptable noise levels are examples of factors affecting 'psychological safety'. The inability to satisfy these needs in the facility setting may not render the facility unusable, but will significantly affect *productivity*.
- *Belonging/social needs.* The necessity to interact with others is not only an organisational need but also fulfils an individual's needs beyond basic physiological and security needs. With the possibility of home-working, the role of the office in supporting this and other higher needs has become more important than ever. In every other respect, tele-working fulfils functional requirements including knowledge-based resources. What a modern office environment can provide is an opportunity to work in a collaborative environment that stimulates trust, understanding and the formation of a common culture. If such work environments do not capitalise on this need for social belonging, sceptics regarding the future of the office may be justified. Fulfilment of this level through the support of the FM environment will enable the realisation of *organisational intelligence* – both on an emotional and cognitive plane.
- *Esteem needs.* Having achieved a sense of inclusion, by involvement with a variety of work groups (both formal and informal teams), individuals will then seek confidence in their own achievements and those of groups with which they associate. These esteem needs relate to issues of respect and should not be confused with status needs, which are all too familiar in the facilities management context (e.g. the desire for more office space or a window aspect). Innovations in branding and the acknowledgement of group identity through workplace design provide mechanisms for developing esteem needs, reinforcing confidence in the individual, their team, their product and their service. Facilities that fail to support this need might be described as *unstable* environments.

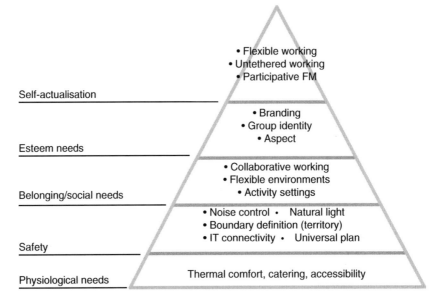

Figure 4.4   Workplace interventions to satisfy the hierarchy of needs (adapted from Maslow, 1943).

- *Self-actualisation.* The pinnacle of the needs hierarchy as identified by Maslow is self-actualisation, which describes a state in which individuals are able to attain all of the things they think they are capable of. As such, an employee can envisage a long-term contribution to be made in an organisation by exercising their innate potential. As indicated in Figure 4.4, innovations in home–life balance, flexible working and on-site crèche facilities are all examples of facilities management interventions that would support this need fulfilment. Emerging practices of participative facilities management and user involvement characterise an approach that supports the fulfilment of such a need.

Maslow's hierarchy of needs forms a bedrock theory for understanding the value-adding role of facilities management and it is perhaps surprising that current FM practice often fails to exploit the insights derived from its use. Some of these insights include:

- An appreciation of the evolution of facilities management, characterised by an early preoccupation with physiological and safety needs eventually being superseded by less tangible concerns related to esteem and self-actualisation. A similar approach has recently been advocated by Urwiler and Frolick (2008) as the basis of a maturity model in the IT sector.
- When used as a maturity model for FM, an appreciation of the differing levels that prevail in different economies, sectors and countries is instructive in deploying an appropriate FM strategy.

- A recognition that certain facilities management innovations can, at the same time, enhance the fulfilment of one 'needs' level and at the same time detract from the fulfilment of other more basic needs. For example, the introduction of untethered 'work anywhere' workplace strategies may enhance self-actualisation and esteem need but at the same time erode security needs.
- Attempts to reduce operating costs typically seek to ensure that *efficiency* is sought through the maintenance of lower-level needs (e.g. functional use of space). However, such measures need to fully recognise the impact on higher-level needs, which are invariably harder to recognise.
- Energy and sustainability impacts on various levels of the hierarchy in relation to thermal comfort (physiological level), lighting (security level), user controls (security and belonging/social interaction level) and access to a view/aspect (self-actualisation).

What is evident from consideration of the hierarchy of needs is the elusive nature of facilities performance measurement. In recessionary periods a preoccupation with efficiency may indeed drive out those characteristics that address effectiveness issues more often associated with higher-level needs. In the age of the knowledge worker, fulfilment of higher-level needs become not just a desirable but an essential feature.

4.7.2   *Two-factor theory*

Linked to the idea of 'needs' is the issue of 'expectation', which was addressed by Herzberg (1964) in his 'two-factor theory', which is now more commonly known as the Motivator–Hygiene theory. Herzberg collected data from interviews with engineers and accountants in the Pittsburgh area during the 1960s. Like Maslow, he wished to know what motivated workers. However, Hertzberg added a significant new dimension to Maslow's original theory by suggesting that a two-factor model of motivation existed. This was based on the idea that a certain set of factors or incentives lead to *dissatisfaction* (or the avoidance of dissatisfaction) whilst another set of entirely different factors gives rise to *satisfaction*. He argued that dissatisfaction and satisfaction did not occur on a simple continuum. The first set of motivational factors were described as 'hygiene' factors (associated with the avoidance of dissatisfaction) and included such things as minimum salary levels, safety and pleasant working conditions, company policy, supervision and technical issues. The absence of any of these factors invariably led to dissatisfaction. In contrast, another set of factors were identified whose absence did not lead to dissatisfaction but were instrumental in making workers happy and satisfied. Factors such as achievement, competency, status, personal wealth and self-realisation were identified in this group of 'motivation' factors. The motivating factors invariably were associated with the nature of the work itself in terms of opportunities for assuming responsibility and recognition.

What are the implications of the motivation–hygiene theory in modern facilities management? Building on the ideas of Maslow, it suggests that in order to improve attitudes towards work and productivity it is necessary for the FM to attend to both hygiene factors and motivator factors. Simply eliminating dissatisfaction by dealing with hygiene factors within the built environment will not ensure that users of space will be satisfied. It is only by understanding the specific needs and role of those working within the space (specifically, the higher-level psychological needs identified by Maslow) can we begin to address satisfaction (as opposed to levels of dissatisfaction).

4.8   FM in the experience economy

Facilities undoubtedly fulfil a radically different role to that of past eras. This is set to continue with what Pine and Gilmore (1998) describe as the 'experience economy'. In our discussion so far we have noted the increasing role of service over product in FM. Moving one step further they suggest that 'as goods and services become commoditised the customer experiences that vendors/sellers create will matter most'. They propose that 'experience is not an amorphous construct, it is as real an offering as any service'. In their analysis they advocate that we should not simply lump together the idea of 'experiences' with services; indeed, they argue that experiences are as fundamentally different from services as services are from goods.

The ideas of Pine and Gilmore (1998) can be traced to the earlier observations of Toffler (1984) who, unlike Gilmore and Pine, argued that the 'experience' element was inseparable, playing an increasingly embedded role in the design of products or services:

> We shall go far beyond any functional necessity, turning the service, whether it is shopping, dining or having ones haircut – into prefabricated experiences (Toffler, 1984: p. 207).

In Pine and Gilmore's analysis they argue that organizations involved with staging experiences should apply a deliberate strategy in designing and promoting an experience so that it is possible to command a fee. The term 'experience' in this regard 'occurs when a company intentionally uses services as the stage and goods as props, to engage individual customers in a way that creates a memorable event' (Pine and Gilmore, 1998: p. 98). One important characteristic of the fourth economic offering presented by the experience economy is that experiences are intrinsically personal, capable of existing only in the minds of the individual such that no two individuals have the same experience. Engagement in an experience involves a physical, intellectual and an emotional commitment in contrast to commodity goods (facilities) and services, which remain external to the buyer.

More and more we see evidence of organisations that use goods (physical assets) as props and services as the stage to enable memorable experiences. This is particularly evident in the retail sector, with shop fit-outs occurring at increasingly shorter time intervals to enable the staging of experiences. We see the advent of 'shoppertainment' typified by organisations such as Apple, Nike and Bose. In the extreme, the shop is no longer simply a place to purchase goods but places where themed experiences occur. Having become immersed in the experience the shopper then proceeds to purchase the product offering online.

Experiences are not confined to the retail sector and form an increasingly important part of many other FM sectors. Therapeutic environments in hospitals (Abbas and Ghazali, 2010), personal learning environments (PLEs) in schools, new ways of working in offices – all require a changing role of the facilities manager as 'stager' rather than service 'provider'. Moreover, the method of delivery may be fundamentally different such that experiences are revealed over a duration rather than delivered on demand, which represents the current service paradigm.

The emergence of the experience economy explains the increasingly concertinaed emphasis on short-life elements and services in FM. As observed by Duffy:

> Add up what happens when capital is invested over a 50-year period: the shell expenditure is overwhelmed by the cumulative financial consequences of three generations of services and 10 generations of scenery (Duffy, 1990: p. 80).

The theatrical analogy used by Pine and Gilmore (1998) mirrors Duffy's (1974) earlier approach, which illustrates the decision-making layers associated with the design and operation of buildings. In just the same way that theatre performances rely on flexible scenery, props and sets (settings), a similar such approach is increasingly observable in modern facilities. The layered model consists of:

Shell → Services → Scenery (space plan) → Settings → Stuff → Soul

The layers refer to (1) the building *shell* with a life of 50–75 years; (2) *services* including heating, ventilation, lighting and cable distribution with a life of between 15 and 20 years; (3) *scenery* relating to fixtures and fittings incorporated at the fit-out stage to match the user requirements; (4) *settings* referring to the day-to-day management of furniture and equipment; (5) *stuff*, here referring to front-line delivery of FM services (servicescape) to customers, not necessarily involving physical artifacts associated with the building; and (6) *soul*, describing the staging of non-routine, memorable and personally distinct experiences.

The addition of the 'stuff' layer by Brand (1994) originates from his original six categories in a similar framework: 'site, structure, skin, services, space plan and stuff'. Significantly, he proposed an additional seventh category of human 'souls' as the most significant stimulus for change. This is akin to the 'experience' offering associated with memorable and personal experiences beyond conventional service delivery and, as such, has been included by the authors in Figure 4.5.

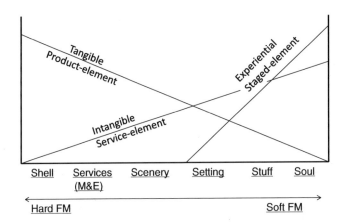

Figure 4.5 The emergence of the experiential element of the building layered model (adapted from Duffy, Laing and Crisp, 1992; reprinted with permission, Emerald Copyright © 1992).

In order to realise the full potential of the experience economy, Gilmore and Pine (1998) suggest five key measures:

- Theme the experience.
- Harmonise the experience with positive cues.
- Eliminate negative cues.
- Mix in memorabilia.
- Engage all five senses.

If services are to be seen as the stage and buildings and their fittings as props to engage building users, each of the five key measures provides a new insight into the role of the facilities manager as 'stager'. The profound changes brought on by the experience economy will provide a new opportunity for facilities management organisations, who currently find that through the process of commoditisation, their offering becomes commonplace and competes on the diminishing returns of price. Contrariwise, the experience economy 'threatens to render irrelevant those who relegate themselves to the diminishing world of goods and services' (Pine and Gilmore, 1998: p. 105).

4.9   Conclusions

This chapter has sought to examine the organisational challenge of facilities management. It is evident that FM strategy does not dictate any one single structural solution, be it outsourcing, bundling or in-house operation. As observed by Peters and Waterman (1982: p. 4), 'strategy rarely seemed to dictate unique structural solutions. Moreover, the crucial problems in strategy were most often those of execution and continuous adaptation: getting it done, staying flexible.' This chapter began by looking at the challenge of performance measurement and its potential misuse. It was suggested that standardisation in this and other areas will allow the widespread commoditisation of facilities management at a global level. However, whilst commoditisation will undoubtedly stimulate growth of the FM outsourcing market, it will also create challenges in terms of increased competition on the basis of price, i.e. the challenge of product differentiation, which can be achieved through other relationship types: namely *customisation, consultation* and *collaboration.*

The changing nature of FM will dictate a varied response to the market. This changing pattern is explained by Maslow's hierarchy of needs, which is as relevant to modern-day work environments as any previous era. Taking this one step further, facilities management will increasingly exploit buildings in order to stage memorable experiences. Flexibility will be central to this approach.

The final part of this chapter looked at the organisational issues capable of supporting an FM environment – one characterised by rapid change and the demands for flexibility. Emergence from purely 'commoditised'-type relationships to ones based on *customisation, consultation* and *collaboration,* demonstrate an increasingly sophisticated FM market. From the client's perspective, flexibility in outsourcing contracts is acknowledged as perhaps the most pressing issue of the time. Sourcing flexibility has various dimensions and prompts several strategic responses. What is increasingly clear is that reliance on an exhaustive contract specification may not be the complete answer for an industry dealing with interdependency and complexity on an unprecedented scale.

Practitioners, having reflected on this chapter, will see that hard FM and soft FM demand differing management approaches. As such, a layered decision-making approach based on an articulation of product, service and experience is necessary in order to capture fully the intangible as well as the tangible effects of FM provision.

References

Abbas, M.Y. and Ghazali, R. (2010) Healing environment of pediatric wards. *Procedia – Social and Behavioral Sciences*, **5**, 948–957.

Blau, P.M. (1964) *Exchange and Power in Social Life.* Transaction Publishers.

Brand, S. (1994) *How Buildings Learn What Happens after They're Built.* Penguin Books, New York.

Cant, M. and Jeynes, L. (1998) What does outsourcing bring you that innovation cannot? How outsourcing is seen – and currently marketed – as a universal panacea. *Total Quality Management and Business Excellence*, **9**(2), 193–201.

Davenport, T.H. (2005) The coming commoditization of processes. *Harvard Business Review*, **83**(6), 100–108.

Duffy, F. (1974) Office design and organizations: 1. Theoretical basis. *Environment and Planning B: Planning and Design*, **1**(1), 105–118.

Duffy, F. (1990) Measuring building performance. *Facilities*, **8**(5), 17–20.

Duffy, F., Laing, A. and Crisp, V. (1992) The responsible workplace. *Facilities*, **10**(11), 9–15.

Gronroos, C. (2000) *Service Management and Marketing*, 2nd edn. John Wiley & Sons, Ltd, Chichester.

Halachmi, A. (2005) Performance measurement is only one way of managing performance. *International Journal of Productivity and Performance Management*, **54**(7), 502–516.

Herzberg, F. (1964) The Motivation-Hygiene Concept and Problems of Manpower, Personnel Administration (January–February), pp. 3–7.

Hui, E.Y.Y. and Tsang, A.H.C. (2004) Sourcing strategies of facilities management. *Journal of Quality in Maintenance Engineering*, **10**(2), 85–92.

Jensen, M.C. and Meckling, W.H. (1976) Theory of the firm: managerial behavior, agency costs and ownership structure. *Journal of Financial Economics*, **3**(4), 305–360.

Kishore, R. *et al.* (2003) A relationship perspective on IT outsourcing. *Communications of the ACM*, **46**(12), 86–92.

Maslow, A. (1943) A theory of human motivation. *Psychological Review*, **50**(4), 370–396.

Maslow, A., Frager, R. and Fadiman, J. (1970) *Motivation and Personality*. Harper & Row, New York.

McGregor, D. (1960) Theory X and theory Y. *Organization Theory*, 358–374.

Peters, T.J. and Waterman, R.H. (1982) *In Search of Excellence: Lessons from American's Best-Run Companies*. Harper & Row, New York.

Pfeffer, J. and Salancik, G.R. (2003) *The External Control of Organizations: A Resource Dependence Perspective*. Stanford University Press.

Pine, B.J. and Gilmore, J.H. (1998) Welcome to the experience economy. *Harvard Business Review*, **76**(4), 97–105.

Poppo, L. and Zenger, T. (2002) Do formal contracts and relational governance function as substitutes or complements? *Strategic Management Journal*, **23**(8), 707–725.

Schumpeter, J.A. (1934) The Theory of Economic Development: An Inquiry into Profits, Capital, Credit, Interest, and the Business Cycle. University of Illinois at Urbana-Champaign's Academy for Entrepreneurial Leadership Historical Research Reference in Entrepreneurship.

Sia, S.K., Koh, C. and Tan, C.X. (2008) Strategic maneuvers for outsourcing flexibility: an empirical assessment. *Decision Sciences*, **39**(3), 407–443.

Toffler, A. (1984) *Future Shock*. Bantam Press.

Urwiler, R. and Frolick, M.N. (2008) The IT value hierarchy: using Maslow's hierarchy of needs as a metaphor for gauging the maturity level of information technology use within competitive organizations. *Information Systems Management*, **25**(1), 83–88.

Williams, B. (1996) Cost-effective facilities management: a practical approach. *Facilities*, **14**, 26–38.

Williamson, O.E. (1985) *The Economic Institutions of Capitalism: Firms, Markets, Relational Contracting*. Free Press, New York.

Zeithaml, V.A. and Bitner, M.J. (2000) *Services Marketing: Integrating Customer Focus across the Firm*, 2nd edn. Irwin McGraw-Hill, Boston, MA.

British Standard

- BS EN 15221-1:2006: Facility management. Terms and definitions
- BS EN 15221-4:2011: Facility management. Taxonomy, classification and structures in facility management
- BS EN 15221-5:2011: Facility management. Guidance on facility management processes
- BS 8536:2010: Facility management briefing. Code of practice
- BS 1192:2007: Collaborative production of architectural, engineering and construction information. Code of practice

5 Managing People Through Change

5.1 Introduction

5.1.1 Scope of the chapter

This chapter examines how the modern-day facilities manager deals with the unprecedented rates of change affecting organisations. In past decades, incremental change has been a familiar phenomenon to most facilities managers. This predictable change gave rise to a level of turbulence and manageable disturbances. The challenge for the FM was to ensure a return to a 'status quo' by carefully forecasting accommodation needs. The modern environment in which the FM professional operates is very different. It is characterised by rapid, major and unpredictable change. This impacts on the systems, styles, structures, staffing and work habits of occupying organisations. As a result, the challenge has become not just facilitating learning through the built environment. Unlearning and relearning is sought through the design and operation of facilities. This chapter explores the idea of the 'service concept', which provides a proactive rather than a reactive retort to uncertainty and unpredictability. Integral to this model is an appreciation of how facilities impact on organisational behaviour. More particularly, how does space impact on the 'psyche' of the organisation? This includes the temporal influence of space. The subsequent discussion examines management principles that can equip the facilities manager with the necessary tools for dealing with the inevitable inertia that accompanies organisational change.

5.1.2 Summary of the different sections

- Section 5.1. Introduction.
- Section 5.2. Examines how changes foisted upon an organisation can lead to attrition and change fatigue. However, when facilities are impacted, such changes also present an opportunity to embrace new service concepts.

Facilities Management: The Dynamics of Excellence, Third Edition. Peter Barrett and Edward Finch.
© 2014 John Wiley & Sons, Ltd. Published 2014 by John Wiley & Sons, Ltd.

- Section 5.3. Articulates a stepwise approach to the evaluation of 'new service concepts'. This model identifies the inherent risks as well as opportunities for adjustment when introducing a new service concept.
- Section 5.4. Challenges the prevailing view of space as a simple physical commodity. It highlights how 'organisational space' is used to express power, establish rootedness, convey new experiences and facilitate unanticipated uses.
- Section 5.5. Explores how rootedness and place attachment affects every organisation. Examines how 'uprooting' building users has repercussions well beyond the physical considerations of move management.
- Section 5.6. Considers proactive approaches for dealing with the 'psyche' of organisations coming to terms with physical change. A case study illustrates how 'grief models' provide new insights into the relocation process.
- Section 5.7. Conclusions.

5.2 Change as a threat or opportunity?

Organisations are in a continual state of flux. Indeed, the very existence of facilities management reflects the growing need of organisations to respond to such upheaval. For many organisations, the changing external and internal environment gives rise to new opportunities – and new services. These new services are not simply FM 'add-ons' to allow the smooth operation of buildings: they sit at the heart of strategic change. For successful FM operations, the driving force is invariably the 'service concept'; such a concept is a prerequisite to any significant transformation. New ways of working, collaborative working, work–life balance, inclusive culture – all represent ideas that can potentially revolutionise how facilities are conceived as part of a wider organisational initiative. Such an approach enables the FM to operate in a proactive manner, seeing change as an opportunity to orchestrate service redesigns. In this chapter, we consider how a proactive approach to change can be guided by a stepwise model based on the service concept at its core. Following on from this, we examine how facilities management interventions are inextricably linked to wider change issues. Rather than being confined to the reapportionment of space, we are forced to examine how space impacts on deep-seated organisational issues. These issues in turn require a careful re-examination of the facilities manager's role in the context of change.

5.3 New service implementation cycle

It is tempting to place the 'facility' at the centre of any change management initiative. A new building or a major refurbishment heralds a physical manifestation of organisational change. The agenda may be driven by property investment decisions, repair and replacement decisions or ongoing attempts to satisfy organisational growth.

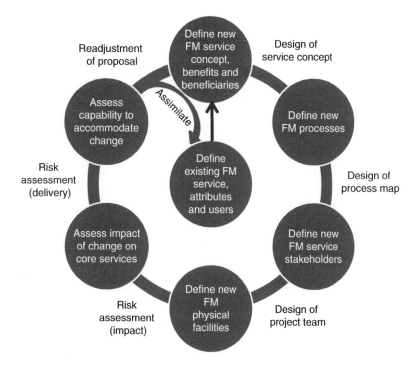

Figure 5.1   A multistage view of change management in FM (adapted from Stuart (1998); reprinted with permission, Emerald Copyright © 1998).

For many proactive organisations, however, new facilities present the opportunity to embed new 'service concepts'. A new facility divorced from any underlying service concept is unlikely to reflect the changing needs of customers, users and stakeholders. With this in mind, we present a change management cycle that uses the 'service concept' as the pivotal element, connecting as it does with the aspirations of the wider business. In FM we think of services as activities that simply enable a facility to function. The reality may be quite different. The service concept may become the *raison d'être* defining how the physical facility is conceived, adapted and used. New service innovations thus serve to justify the upheaval brought about by a new facility.

Figure 5.1 illustrates a multistage analysis of change management in FM based on a cyclical model described by Stuart (1998), which has at its core the introduction of a new service.

If we think of FM initiatives as being service enhancements involving a new service concept (including the redesign of physical assets), it is possible to identify seven key implementation stages:

1. Definition of the proposed **new** service in terms of 'service concept', including beneficiaries and benefits.
2. Comparison with the **existing** service, its attributes and users.
3. Analysis of FM **processes** required to fulfil the new service concept.

4. Identification of the FM team, user groups and the wider implementation team (**stakeholders**).
5. Definition of the requisite **physical** facilities.
6. Assessment of the change **impact** on core services.
7. Assessment of **capacity/capability** when implementing a new service concept.

5.3.1 Stage 1. Defining the new 'service concept'

The term 'service concept' describes the way in which an organisation would like its services to be perceived by its customers, employees, shareholders and lenders. As such it represents a 'mental picture'. Some authors describe this as the combining of separate components into a service package (place, process, people). Zeithaml, Bitner and Gremler Dwayne (1996), in their 'Seven Ps' approach, identify several key elements that go to make up the service offering in its entirety:

- Product
- Process
- Place
- Physical evidence
- People
- Price
- Promotion

When viewing the service concept as a combination of elements, the service package incorporates not only the goods that are used in the delivery of a service but also the environment in which the goods and services are consumed (importantly, the FM environment).

Clark, Johnson and Shulver (2000), however, caution against a 'bits and pieces' approach to the service concept that is common in operations management and marketing. They suggest that the service concept represents much more than simply the DNA of the services or the elements present in a service package. They go on to argue that 'a day at Disney's Magic Kingdom is more likely to be defined by its designers and its visitors as a magical experience rather than as six rides and a hamburger in a clean park' (p. 72). Another illustration of a singular pervasive mental image is given in a quote by Charles Revson, creator of the cosmetic firm Revlon (cited by Kotler, 1997: p. 445), 'In a factory we make cosmetics and in the drugstore we sell hope.' The task of articulating the service concept presents a fundamental challenge to facilities managers who are perhaps more comfortable with the specification of buildings and fit-out.

5.3.2 Stage 2. Comparing with the existing service

The introduction of a new service has implications both for the beneficiaries of the new service as well as those that depend upon the existing service (including existing facilities). Being able to characterise these users and understand how they might be affected by the introduction of a new service is key. It will highlight how existing users might exhibit levels of change resistance and from a pragmatic point of view might be affected by the reallocation of resources in favour of the new project/service.

5.3.3 Stage 3. Defining processes

The routine activities that take place within a facility can be represented in terms of one or more processes. Increasingly, particularly in the manufacturing industries, the envisaged processes dictate the layout and configuration of a facility. In more recent times process mapping is also apparent in organisations focused on service delivery. One clear example is the health industry and hospital planning. By mapping activities and processes it is possible to improve facility layout and design. Furthermore, the possibility of reducing waste becomes more apparent in such an analysis.

The widespread use of Lean Thinking, both in construction and facilities management, illustrates the growing importance of process. Underlying this analytical approach is the recognition of three areas of process improvement (expressed as Japanese terms from which the technique arises): muda (waste or futility), mura (unevenness) and muri (overburden). Muda refers to seven key areas of waste, including transportation, inventory, motion, waiting, overprocessing, overproduction and defects. Traditionally, much of the focus in the manufacturing sector has been on the elimination of waste (muda) and process control. However, when applied to facility design in the service sector, the concepts of mura and muri offer equally expansive opportunities to improve processes.

5.3.4 Stage 4. Identifying the stakeholders

This stage involves identifying a range of individuals and groups that will be impacted by the change, including those that will be instrumental in enabling the change. The issue of stakeholders is addressed in more detail in Chapter 3. From a change management perspective, stakeholders present a particular challenge. Their perceptions and commitment to any proposed change may ultimately dictate the success or failure of the change. Reluctance to commit to a change may be based on legitimate concerns. For the change manager, the challenge then becomes one of understanding their concerns and incorporating refinements where necessary. Five principal components have been identified by Armenakis and Harris (2002) that are known to affect stakeholders' willingness to accept the proposed change:

- Discrepancy (Is there a problem/opportunity to start with? Is a change necessary?)
- Appropriateness (Does the new proposed service overcome the perceived discrepancy?)

- Efficacy (Is this the best way of tackling the change and is the stakeholder/organisation capable of executing the change effectively?)
- Principal support (Can the stakeholder expect support from senior managers?)
- Valence (Is there a perception of personal benefit arising from the change?)

Only by understanding which of the five components influence the perception of the particular stakeholder is it possible to formulate an appropriate change message. A method for conveying appropriate change management messages is described by Hammond *et al.* (2011) based on the five components identified above. Using action research, they were able to illustrate how 'change cynics' could subsequently be used as the principal participants in a change development programme.

5.3.5 Stage 5. Identifying the facilities

Having identified the characteristics of the new service, the processes involved and the people affected by the change, it is possible to explore the physical 'hardware' necessary to support the introduction of the new service concept. This may entail acquiring new leased space, procurement of owner-occupied space, reconfiguration of space, upgrading of existing space or regional redistribution/centralisation of space. Whilst it is tempting to approach this stage from a property asset perspective, organisations are increasingly mindful of the operational impact. As such, this stage is integrally linked to issues identified in previous stages (e.g. change readiness of stakeholders, ability to accommodate processes and ability to reflect the new service concept).

5.3.6 Stage 6. Assessing the impact on core services

The introduction of a new service concept inevitably impacts on core services. This includes both the short-term disruption and the long-term benefits. Repeated changes present the risk of 'change fatigue', so careful consideration needs to be given to any enforced changes. The prospect of moving into a new building may not be as enticing to employees as the facilities manager might anticipate. For external customers, business continuity is paramount and the realisation of long-term benefits should not come at the cost of short-term loss of operational capability.

5.3.7 Stage 7. Assessing the capability to accommodate change

The difficulty of accommodating change is one of the reasons why many organisations postpone changes long after discrepancies become apparent. A change of service concept asks fundamental questions about the capacity of employees to adapt to change. Often, from a facilities management perspective, we focus on the logistical implications involved with a change. This may entail changes in

Service Level Agreements, changes in supplier, changes in 'hardware' and changes in FM personnel. However, much more challenging for the facilities manager in the long term is the ability of building users (individuals and teams) to adapt to their new circumstance. Lewin (1951) identified a three-step model for change that reflects how organisations come to terms with change. This force-field model assumes that organisations seek a status quo that is achieved by the balancing of driving and restraining forces. When discrepancies or mismatches with the external environment become intolerable and threaten the status quo, organisations then have to increase driving forces and reduce restraining forces. To enable the required new behaviour, the equilibrium itself has to be destabilised (unfrozen). Following on from this first step of unfreezing, the second step involves enabling groups and individuals to move towards the desired behaviour. Finally, in the third step, known as the 'refreezing' stage, stability is achieved again. This organisational refreezing involves cultural norms, policies, practices and many other aspects of organisational culture.

If the facilities management team is to implement a successful change operation, they must inevitably address the wider implications for the core business (including the human impact). The above planning stages for evaluating and revising a service concept depend not only on the 'change readiness' of the FM team but also the building users. By carefully profiling the stakeholders and anticipating/responding to their concerns and ideas, the likelihood of a successful change management process can be significantly increased. The relationship between the FM team and the Human Resources team will play a pivotal part in a message conveyance strategy facilitating the change process.

5.4 Multi-faceted 'space'

Buildings are imbued with meaning. Individuals and teams, from the day that they occupy the building space, invest emotion and meaning. Space is then no longer a 'fungible' commodity. In other words, it is not possible to simply substitute one square metre with another in the way that we can substitute crude oil, shares or precious metals.

Facilities managers deal with the commodity of 'space' rather than 'buildings', which is the domain of real estate professionals. An organisation's space requirements can be met by means of novel approaches such as shorter lease terms, sale and leaseback, fully serviced offices and Public–Private Partnerships (PPP), all of which may provide agility in an industry characterised by inertia – the inertia that arises from property ownership. In the words of William James (American philosopher 1842–1910), 'lives based on having are less free than lives based either on doing or on being' and much the same could be said about attitudes to property ownership. Space as a fungible commodity can be easily measured in square metres, easily priced and easily exchanged. However, this view of space that dominates the facilities management ethos brings with it its own problems. In order to address these problems fully we need to visualise space from a number of organisational perspectives. It is only by understanding these alternative perspectives that we can start to address the consequences of an organisational environment in which place ceases to possess any long-term meaning for the individual.

Much is spoken about the need to align facilities management with the strategic objectives of an organisation. However, the language of facilities management is often at odds with that of such organisations. A preoccupation with space as a cost item measurable in terms of square metres demeans the true value of space. Echoing this concern over the narrow interpretation of space is that of Foucault (1980: p. 70):

> ... for generations in the social sciences space was treated as the dead, the fixed, the undialectical, the immobile. Time, on the contrary, was richness, fecundity, life, dialectic.

An important premise in this discussion is that time and space are increasingly the focus of workplace design and facilities management. In relation to organisational research, the time element plays an increasing part. As Taylor and Spicer (2007) suggest, 'until recently the spaces and places that management happens in and through have been portrayed as neutral settings'. They go on to suggest that 'the established subfields of management research tend to "see" spaces as specific commonsense categories that can be separated out from each other empirically and analytically'.

Space can be expressed in many different ways in an organisational setting. Common expressions used include: place, region, locale, workspace, work environments, private/public space, territory and buildings, to name but a few. This variation in the umbrella term associated with space reveals variation in underlying assumptions about the nature of space (often at odds with the assumptions of physical space).

In the next few sections we suggest that, in dealing with change management, a broader perspective on the nature of space is required in facilities management. As such, organisational spaces can be classed in relation to four categories (based on the categorisation of Taylor and Spicer, 2007, and Lefebvre, 1991): space as distance (physical space), space as the materialisation of power relations, space as opportunity and space as experience. By such categorisation we are able to more fully understand the link between organisational time and space and, as a consequence, the challenge that is change management. Figure 5.2 illustrates the four different ways of understanding space expressed in four different quadrants, encompassing:

- Space as distance (evaluated in terms of observed behaviour)
- Space as a means of control (evaluated in terms of political/organisational influence)
- Space as experience (evaluated in terms of affect/sensation)
- Space as opportunity/affordance (evaluated in terms of flexibility for use)

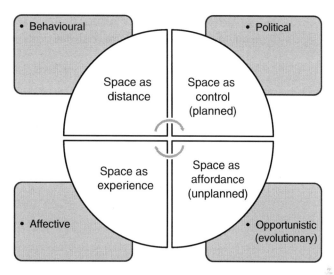

Figure 5.2   Understanding organisational space.

5.4.1   *Space as distance*

Undoubtedly the most common approach to understanding space in facilities management is to consider space in terms of an area or a physical distance. This approach is based on Euclidean geometry, which identifies space as the distance between two or more points. One repercussion of this approach is that spatial distances are amenable to objective assessment in terms of proximity and distance. Such an analysis gives rise to easily measured results acquired using a variety of space planning tools. Post-occupancy evaluation, which can be defined as '... the process of evaluating buildings in a systematic and rigorous manner after they have been built and occupied for some time' (Preiser, 1988), makes extensive use of results based on physical distance. Examples of such techniques include:

- Proxemics. The term 'proxemics' was originally coined by E.C. Hall (1963) who was interested in people's use of 'personal space' in contrast to 'fixed' and 'semi-fixed' spaces. The 'fixed' features in an office, for example, would include walls, doors and partitions, which present immovable boundaries. 'Semi-fixed' feature space, in contrast, would include movable boundaries such as furniture and flexible furniture systems. Finally, of particular interest was the 'informal space', which is characterised by a particular buffer zone that surrounds individuals and varies according to the individual and circumstance. In the study conducted by Hall on the nature of informal space four categories were identified: the *intimate distance*, which applies to behaviours such as embracing and whispering (0.15–0.5 m), the *personal distance*, which applies to behaviours such as conversation among friends (0.5–1.2 m), the *social distance*, which applies to conversations involving acquaintances

(1.2–3.5 m), and distances used for public speaking (3.5 m or greater). The significance of proxemics relates not only to the way people interact with others in daily life but also the organisation of spaces in buildings.

- Affinity matrices. This planning technique seeks to identify relationships and interrelationships that arise between different activities. The intention is to identify a plan that optimises functional requirements. Typically, such an analysis involves a two-step process. The first of these involves the assignment of values relating to the desirability of having one area immediately adjacent to another. Based on these values, it is then possible to establish the most effective plan layout. The second step in the development of an affinity matrix involves the identification of incompatibilities between spaces. Thus, in hospital ward design, proximity between patient beds and the nurse station may give rise to breaches in confidentiality. Similarly, other environmental factors such as fumes, heat and light may give rise to incompatibilities between two types of space. The resulting affinity matrix that considers adjacencies and incompatibilities then allows the identification of an optimal floor layout.
- Behavioural mapping. This technique tracks behaviour over space and time. Two alternative approaches are commonly used: (1) the *place-centred* mapping technique, which focuses on a particular place, or (2) the *individual-centred* mapping, which is based on an individual's movements (Zeisel, 1984). A place-centred map is designed to reveal how or when a particular space is being used, or not used. In contrast, an individual-centred map provides an overview of the daily routine of an individual (e.g. an office worker, a patient, a student).

The advent of advanced tracking technology (such as radio frequency identification, near-field communication (NFC) and global positioning systems (GPSs)) now provides a much richer and more reliable source of information for such studies that focus on behavioural (visible) evidence relating to space usage. Data can be collected that pinpoints the location of individuals at regular intervals throughout the period of study. Both computer-aided facilities management (CAFM) and geographic information systems (GISs) can be used to aggregate such data and can be applied at many different levels from the micro level (workplace) to the macro level (infrastructure management).

When looking at the micro level of workplace layout, it is possible to identify particular types of behaviour and interaction. These have been studied by authors such as Becker (1981), Bitner (1992), Duffy (1997) and Sundstrom, Burt and Kamp (1980). In particular, the adoption of the Open Plan office has been the subject of close scrutiny. Whilst some authors have highlighted the opportunities of flexible space enabled by the use of the Open Plan office design (Duffy, 1997), others have pointed to the detrimental effects arising from the removal of physical barriers. Such detrimental effects include decreased satisfaction, decreased levels of interaction and decreased motivation (Hatch, 1987).

Common to all of the studies discussed so far is the role of the physical arrangement with specific consideration regarding the physical distance between employees, objects, partitions and various architectural/interior design features. Undoubtedly use of studies expressed in terms of basic linear elements has provided significant insight into workplace behaviour as well as opportunities for reducing space and costs based on utilisation rates. These studies have allowed important insights regarding the impact of space on organisational activity. In particular, they illustrate how variations in the distance and proximity of people and resources can significantly affect workplace efficiency, employee satisfaction and workplace loyalty.

However, as pointed out by Taylor and Spicer (2007), the approach also brings with it certain limitations. The first of these is the difficulty of establishing what goes on in the minds of those employees whose behaviour is being observed. Specifically, there is no clear understanding of how employees' perceptions or experiences relate to distance and proximity. Reliance on physical measurements relating to spatiality does not consider how patterns of power and resistance ultimately shape what is observed in terms of distance and proximity. The objection is that the physical layout of built facilities in relation to groups and architectural artefacts does not provide sufficient information about how spaciality is practised.

As noted by Becker (1981), there remains a lack of understanding between those involved with the technical manipulation of organisational spaces (i.e. facilities managers, architects and interior designers) and those involved in the social organisation of the spaces. Such factors include employees, managers and customers. Inevitably this gives rise to conflicts between those that design and those that inhabit the space. This may result in reduced space performance and the emergence of neglected space. This conflict invariably arises from a preoccupation with space as distance, overlooking the other three dimensions of space that will be discussed (space as control/power, space as experience and space as affordance). Advances in information technology may also exacerbate this tunnel vision. Building information modelling (BIM), with its focus on translating the original 'as-built' details into an operational tool, may present a bias towards the technical Euclidean perspective.

5.4.2 *Space as control*

Another way of conceptualising space is in terms of organisational power. This contrasts with the much more 'visible' characteristics of space as physical distance. As argued by Allen (1977) and Taylor and Spicer (2007), the conceptualisation of *space as distance* provides only a partial understanding of space and its use. As such, it explains *how* spaces are used but does not explain *why* spaces are used in the way they are. In order to provide such an explanation, both authors identify *materialisation of power relations* as a key way of understanding economic space.

Early manifestations of the Open Plan environment represented an attempt by supervisors and overseers to carry out surveillance of office worker behaviour to ensure that employees conformed to the prevailing office codes (Markus, 1993). This mimicked that of the industrial situation in what Jay (1986) calls the 'Empire of the gaze'. These spaces were designed to enable easy visual surveillance and control. Modern-day environments are less concerned with such 'powerscapes', largely because observable behaviour has largely been supplanted by 'virtual behaviour'. This has given rise to more abstract managerial systems relating to disciplinary authority and control (Rose, 1990).

Many commentators have argued that building architecture, workplace design and working environments remain pivotal in shaping relations of power. Moreover, the advent of more flexible space has enabled improvised expressions of power relations (both espoused and informal), enabling users to have some say as to the arrangement of flexible furniture systems in a localised workplace environment. The adaptability of modern spaces enables such responsive improvisations. The formation of team areas and group spaces are examples of impromptu settings that fulfil short-term management desires to control behaviours amongst employees.

Associated with the idea of power relations in the workplace are alternative models of facilities management involving greater user involvement. The greatest advocates of such an approach include Friday (2003), who espoused the use of organisation development (OD) in facilities management, and Vischer (2008), who recommends a user-centred approach to facilities management.

Mapping out where the modern-day facilities manager sits in this apparent workplace power struggle is challenging. One enduring leadership model that provides guidance in such a decision-making setting, where one size does not fit all, is the 'general management leadership behaviour theory'. This theory, proposed by Tannenbaum and Schmidt (1973), can be used to identify appropriate leadership styles of the facilities manager when dealing with different FM problems and user groups.

The leadership model of Tannenbaum and Schmidt (1973) presents a continuum of choices for involving employees in decisions, ranging from 'telling' to 'devolving' decisions. Indeed, this continuum is often described as the 'sell/tell' continuum but in fact covers a much wider range of approaches. This continuum is described for decision-making scenarios involving the space planner/facilities manager and building users, as expressed by the seven levels of selling/telling shown below:

1. Architect/FM announces workplace solution.
2. Architect/FM 'sells' workplace solution.
3. Architect/FM presents ideas and invites questions.
4. Architect/FM presents tentative decision subject to change.
5. FM presents problem, gets suggestions, confirms design solution.
6. FM presents problem, gets suggestions within defined limits and asks group to make decisions.
7. FM permits employees to adapt space within limits defined by their team leader.

In certain situations a leadership style that is autocratic (level 1) may be required. This might be driven by concerns over compliance and timeliness in decision making. At the other end of the continuum (level 7) a fully participative decision-making approach may be more appropriate. The opportunity for workplace occupants to engage in FM decision making has increased significantly with the advent of flexible components. For the facilities manager such modifications may be positive, allowing users themselves to personalise workplace design and group settings. However, it may also present problems in terms of sanctioning changes agreeable to all users that are not in contravention of health and safety regulations (e.g. obstruction of walkways in relation to fire regulations).

Territory and territoriality might be construed as subsets of the wide-reaching concept of 'power relations'. We are all too familiar with attempts by organisational units to 'grab' space that appears to be empty, with the result that free space never becomes available to the facilities manager. Such encroachments and 'empire building' related to space are all too frequent, giving rise to inefficient and ineffective space usage. However, as argued by Brown, Lawrence and Robinson (2005: p. 579), 'the motivations for territorial behaviours are far more varied than simply the desire for influence and strategic advantage. The search for personal efficacy may be quite separate from the ability to influence others.' Brown goes on to suggest that the relationship of territoriality to power and politics in organisations is a complex one because they are related concepts but also clearly distinct from one another. Power in organisations originates from attempts to control scarce, valuable resources and the ability to exploit that control through various strategies.

The issue of power relations has become increasingly significant with the advent of home working. The arrival of new information technologies and new management approaches have brought about a blurring of boundaries between the work environment and the home (Perlow, 1998). Increasingly we see that activities attributed to the workplace now become part of home and leisure existence. Equally, those sentiments normally associated with a home are incorporated with the work environment. This has given rise to the 'boundaryless' organisation.

The work of Taylor and Spicer (2007) suggests that organisational spaces provide a mechanism of control that is widely used by management, designers and other planners of space. They contend that spatial configuration may be driven by a desire for surveillance of workers and as a means of dissolving the home and work boundaries.

5.4.3 Space as experience

Our experience of modern spatial environments is increasingly transient. Whether we are talking about office spaces, hospital wards, school classrooms or shopping centres, we are ever more aware of the impermanence of space. This sense of impermanence is more acute and intimate than at any time in history.

As facilities managers we are often instrumental in accelerating transience. We respond to organisational flux and turbulence by reconfiguring relationships. These relationships may involve things, places, people, organisations and ideas. The relationship with things become increasingly compressed, foreshortened and collapsed in time. Space management tools allow us to locate and relocate individuals and groups without reference to the impending consequences.

The single greatest indicator of transience in office environments is the 'churn' rate. Churn can be defined as the number of employees moved in a year expressed as a percentage of the total number of office employees. Typical churn rates in modern office environments are around 40%, although churn rates as high as 150% are not uncommon. Thus, if an organization has 1000 occupants on site and a churn rate of 50%, this equates to moving 500 people every year. Based on a cost of £500, that is an expense of £250,000 per year. We might also include the indirect costs associated with the loss of productivity to an organisation. Invariably this is measured based on downtime: in other words, the disruption time when employees are unable to carry out normal activities including working with clients. In such a calculation, loss of productivity is often quoted in 'hours'. However, the true cost of moves and relocations dwarf this figure. Transience brings with it a human cost that is far more insidious. Perhaps a more appropriate definition of churn is 'to agitate or produce violent motion', which captures the sentiment that many office workers have about frequent relocation and change.

Transience is not only associated with the negative aspects of disruption. It also offers the possibility of continuous novelty and newness. This pursuit of 'experience', which traditionally has been associated with consumer products and more recently with event management, has now entered the dictionary of facilities management. This phenomena was foretold by Toffler (1970) in his influential book, *Future Shock*, which describes '... a shift that will lead to the next forward movement of the economy, the growth of the strange new sector based on what can only be called the "experience industries"' (Toffler, Toffler, 1970: p. 204). The space that we design and manage is increasingly based on psychological needs rather than any utilitarian needs in the ordinary sense. Spaces become stages. Rather than fulfilling a fixed function, spaces provide the backdrop for a continuously changing landscape. Thus, the facilities manager becomes responsible for what Toffler describes as 'experiential production'.

One example showing how the psychologisation of space has progressed since the original coining of the phrase by Toffler some 40 years ago is described in the Inset, which relates to the roll-out of the JobCentre Plus design across more than 1000 locations in the UK. The work of lead architects Lewis & Hickey covered a four-year period, working with CiD who were employed as the brand consultants for the design development and implementation of Jobcentre Plus. The brand consultants CiD were responsible for rationalising the identity, as well as creating and refining brand components.

JobCentre Plus: An 'engineered' environment

Walk into any branch of JobCentre Plus in the UK and one is immediately struck by the sophistication of the engineered environment. Designed to enable jobseekers and those seeking benefits to go about their business, it is a world away from the kind of environment encountered two decades previously. The old environment comprised national offices that were austere and often brutal in appearance. Harsh fluorescent lighting, physical barriers in the form of protective screens between members of the public and the service provider, rudimentary seating in overpopulated and serried ranks, conspired to produce an environment of confrontation and lack of self-worth.

The modern environment addresses the 'psyche' as well as the functional requirements of the user. Indeed, behaviour is influenced by subtle rather than obvious means. The jobseeker no longer becomes a passive recipient in the service exchange. The environment is designed to be welcoming and familiar. The two-colour carpeting subtly conveys to the visitor the zonal separation of public space and private space and directions of travel. This is reinforced by the differing lighting schemes that provide zoned lighting. The space presents a journey rather than a single interaction. Customers may be trying to obtain information on jobs, others may be seeking financial support, yet others may be seeking personalised advice. Each of these requirements demands a different spatial experience. For the autonomous jobseeker, access to a freestanding kiosk that allows them to interrogate current availability of jobs may be sufficient. For others, a much more involved and personalised interaction with an adviser may be required. Some may be in need of financial support.

Not only do their requirements differ but their predicament may also differ. Some may be agitated, confrontational or confused. Some may be accompanied by children or other family members. Accommodating a variety of requirements in a 'one-stop shop' environment is challenging. The interior space has to provide 'intelligence' about how to negotiate the various service interactions and match these with the needs of the customer.

Greeting and security staff need to reinforce the messages of the designed space. This may vary from simply directing customers to the correct location; for others, it may be to do with ensuring privacy and avoidance of confrontation in a public space. Sensitive issues such as disputes over payment entitlement may require a 'back door' solution.

What is clear from this modern day 'multiple activity setting' is that the complex interactions that arise depend upon the use of subtle cues in the environment. Without such cues the service encounter becomes unintelligible, particularly for infrequent customers. Only by 'engineering' a complex yet apparently simple servicescape is it possible to effectively control such varied interactions.

5.4.4 *Space as opportunity*

So far we have explored three differing ways in which organisations perceive and manage space. The first of these, space as distance, accords with our traditional approach to analysing space usage in facilities management. We also considered another dimension of space, which provides a different insight into how space is designed, allocated, procured and used: that of political space (space as control). We saw that, with the advent of flexible workplace solutions, interventions motivated by political intentions have become not only part of the original design solution but also part of the day-to-day management solution. Another way of looking at space is from the experiential perspective. This is the third way of viewing organisational space, which, like political space, confounds our traditional way of looking at space from the facilities management perspective. Both the political perspective and the experiential (what is described in psychology as the affective dimension relating to human emotions) provide a deeper understanding of how space is used and can be used. Indeed, looking at the problem through these two lenses enables the facilities manager to understand the *why* as well as the *how* that underpins any space planning strategy. The final way of looking at space that perhaps challenges conventional ways of thinking of space is that of 'affordance'. Space is invariably used in a manner that was not originally intended. User behaviour is often at odds with the original intentions of the designer. Likewise the adaptations made by the facilities manager (based on pragmatic day-to-day considerations) may be at odds with the original design. Consistently we see building users using space for entirely different reasons to that originally conceived. This may be evident in changes sanctioned by management or simply by individual unsanctioned actions. Should such changes be encouraged or challenged? This is an ongoing question that confronts the facilities manager. An understanding of the opportunistic dimensional space usage provides some answers to this dilemma. It challenges our traditional design and planning approach that often makes many assumptions about the changing needs of users.

Affordance can be thought of as a three-way relationship between animal, environment and action. The idea of 'affordances' was originally conceived by Gibson (1979), who defined the term affordance in the following way:

> The affordances of the environment are what it *offers* the animal, what it *provides* or *furnishes*, either for good or ill. The verb *to afford* is found in the dictionary, but the noun *affordance* is not. I have made it up. I mean by it something that refers to both the environment and the animal in a way that no existing term does. It implies the complementarity of the animal and the environment (Gibson, 1979: p. 14).

An example of an affordance in the office environment would be an object that has a rigid, flat, level and extended service that appears at knee height to the human observer. If this precise combination of properties exists, then it affords 'sitting on'. Provided that visual information is available regarding these four properties and such information is detected, then the affordance of 'sit-ability' can be perceived. Notice how this differs from the designation of 'seats', which involves a pre-engineered solution to which the user needs to conform in terms of behaviour. In practice, users are much more opportunistic in terms of how they choose to use space and objects within spaces. They make use of latent capabilities. As noted by Zeisel (1984: p. 132):

> Objects imply obvious options for use: seats in telephone booths are for sitting down when calling, bathroom sinks for washing hands. At the same time they have a host of less obvious latent implications limited only by users' physical capabilities, daring, and imagination. The telephone seat provides tired non-callers a place to rest. Sinks in school bathrooms often fall off the wall because they are sat on by teenagers taking a cigarette break between classes. On a hot summer day urban fountains turn into swimming pools. These objects can be seen as props for behavior.

As noted by Gibson (1979) the theory of affordances provides a new theoretical basis for design. Much could be said about the basis for a new approach to facilities management. In Gibson's view, products are not simply a patchwork of forms to users but represent 'the possibilities of actions'. Thus, by interacting with the product (e.g. a room) the user is able to initiate a sequence of possible actions to achieve their specific goal.

Not all affordances are desirable. Thus, for example, pavement furniture such as benches or stairs may provide the affordance of 'skateboard-ability', thus attracting unwelcome visitors outside prestige offices. As Zeisel (1984) suggests, the scope of affordances are only limited by the physical capabilities, daring and imagination of the individual.

The opportunistic or affordance-based approach to the design of space challenges conventional thinking regarding 'usability'. Often, the steps used in determining usability are concerned with:

- making it more efficient to use (taking less time to accomplish a particular task);
- making it easier to learn (the operation can be learned by observing the object);
- making it more satisfying to use.

However, the assumption in the 'usability' paradigm is always that the use requirement is prescribed. In other words, the facilities manager knows for what purpose the space or the workplace system is to be used. In practice, users may invariably seek the fulfilment of unanticipated needs.

The linkage between affordance, opportunistic behaviour and user-centric facilities management is evident (Holm, 2006; Vischer, 2008). With the user-centred design approach, the product is designed with its intended users in mind at all times. In what is known as the 'participatory design' paradigm, users themselves become actual members of the design team.

Understanding the performance of a building involves much more than a simple analysis of the physical entity expressed in terms of coordinates and specifications. Building use entails a much richer and more complex temporal interaction that is intimately linked to the time dimension and organisational dynamics. In the subsequent discussion we look at how long-term connections over time impact on users' attitudes to facilities, space and the workplace. Workspace becomes a 'sticky' entity rather than a 'non-stick' commodity. This observation has profound implications for the FM's approach to churn, adaptation and relocation and their evaluation of resulting costs.

5.5 Relinquishing the 'old'

The capacity of building occupants to embrace a new environment (be it a school, office, shop or other setting) is pivotally influenced by their attachment to the more familiar 'old' facility. This often deep-felt connection known as 'place attachment' is defined as one's emotional or affective ties to a place, and is generally thought to be the result of a long-term connection with a certain environment (Altman and Low, 1992). This phenomenon is fundamentally distinct from the initial emotional response to an environment that reflects its aesthetic qualities. This sense of 'rootedness' is disrupted when we lose our familiar place. As observed by Jeffreys (1995: p. 44):

> For many employees it is as if they have slipped through a crack in the universe and no longer recognize where or who they are. What was once a secure home-away-from-home has become a frightening, unfriendly and even hostile workplace.

Jeffreys (1995) goes on to argue that embracing the 'new' is not the only problem, but leaving the old culture behind in what he describes as workplace grief.

Milligan (2003) extends the idea of organisational death by arguing that certain organisational site moves (in particular, those in which employees hold a strong place attachment to the environment they are leaving) come to represent a form of organisational death. As a result of the study, several principles were identified to address the concerns of those affected by a move (and dealing with the resulting grief):

1. Recognition of employee feelings of loss and legitimization of employee nostalgia for the old site.
2. Better managerial communication with employees prior to and after the move.
3. The holding of parting ceremonies at the time of the move.
4. Memorialising and preserving artefacts from the old site.
5. Reinforcing and/or re-establishing organisational patterns and rules disrupted due to the move.

5.6   Embracing the 'new'

The replacement of old habits, practices and ways of doing things is a necessary, if often uncomfortable, process. Moreover, it is intractably linked to the challenges of the new physical setting. Some insight into this process can be found in the broader literature relating to experiential learning. Foremost in this respect is that of Gestalt psychology (Nevis, 1987), which suggests that day-to-day incidents and moments tend to move into the foreground of a person's attention, which in turn prompts individuals to act on these incidents. This is then succeeded by a final process of resolution. Applying the Gestalt perspective, a sequence of sensations can be identified, which is described as the 'cycle of experiencing'. This sequence takes the form:

$$\text{Sensation} \rightarrow \text{Awareness} \rightarrow \text{Energy Mobilisation} \rightarrow \text{Contact} \rightarrow$$
$$\text{Resolution} \rightarrow \text{Withdrawal of attention}$$

In the context of relocations and churn, such stages might be associated with (a) alerting employees of the need for the move; (b) engaging in dialogue regarding the nature and implementation of the move (pre-occupancy evaluation); (c) mobilising activities (including planning and the physical move itself); (d) the encounter with the new environment; (e) resolution either by a planned process of *post-occupancy evaluation* or unplanned 'fixing to suit'; (f) a period of stabilization when scrutiny of the move consequences are no longer required. In reality, withdrawal of attention may never occur in the face of continued changes that typify modern work environments.

How do building users let go of the old and accept the new? Various modern-day change management writers have suggested that such a process involves multiple stages. Bypassing these stages by enforcing change only serves to delay an inevitable emotional voyage that needs to be undergone (both at the individual and organisational level). Most understanding of the change process derives from the seminal work of Elisabeth Kübler-Ross (Kübler-Ross, Wessler and Avioli, Kubler-Ross, Wessler and Avioli, 1972) on bereavement, in what is now referred to as the 'five stages of grief' model.

In the 'five stages of grief' model, concerned with coping with bereavement, she interviewed more than 500 dying patients. The model, based on a grounded approach, identified discrete stages that describe how people deal with grief and tragedy. The model itself identifies several phases through which individuals pass:

1. Denial
2. Anger
3. Bargaining
4. Depression and
5. Acceptance.

The 'five stages of grief' model has today received much wider attention as a basis for understanding organisational change as well as individual changes involving upheaval or trauma. Indeed, the phenomenon of grieving has been largely

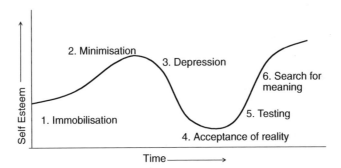

Figure 5.3   Transition model (Reprinted with permission of John Wiley & Sons, Ltd © 1976, from Adams, Hayes and Hopson, 1976: Figure 1.2, p. 13).

acknowledged as an important phenomenon, even when there is the promise of improvement (i.e. new facilities, new functionalities). The work of Elrod and Tippett (2002) extends the 'five stages of grief' model to *organisational* change and shows experimentally that the performance 'dip' associated with the implementation of organisational change is measurable. Two survey instruments were used to assess the performance of diverse multi-disciplinary teams in the studies. Another such model is that of Adams, Hayes and Hopson (1976) (see Figure 5.3), which presents a six-stage model of change and organisational transition.

A recent case study funded by the Leadership Foundation in Higher Education (UK) was undertaken between 2011 and 2012 and sought to identify the discrete transitions of managers affected by relocations. This study was based on a constructivist approach using a modern-day adaptation of the Kübler-Ross five-stage model. The objective of the research was to evaluate the transitional stages of individuals and working groups affected by a relocation from a conventional facility to an 'incubator'-type high-tech learning environment.

MediaCity at the University of Salford (UoS) provided the live case study for the research. Occupied in September 2011, it provides 9600 m² of integrated space comprising 2800 m² of open public access spaces including 'living laboratories' and large public performance spaces. This is juxtaposed with high-end teaching and research facilities developed and designed as a flexible drop-in facility. The space is designed to facilitate an incubation culture where cross-fertilisation of education and learning can occur in a natural organic way, with the intention of creating complex outcomes.

The visionary aspirations of Mediacity bring with it many human and organisational challenges in common with many higher education (HE) institutions. Whilst the physical move can be legislated for, the 'psychological move' required by staff and students to realise the potential of a new working environment was more challenging. The relocation involved all aspects of change including the breaking of emotional ties, the introduction of new processes (enabled by digital technologies) and familiarisation with a new facility. For managers, it entailed being a pathfinder and scout, as well as guiding the progress of others through the transition. The case study analysis of MediaCityUK sought to examine the evidence regarding the emotional changes associated with the move.

The tool used in the study was based on 'personal construct psychology' (PCP). Founded on the work of Kelly (1955), the technique assumes that people act like scientists: they construe the world by putting personal experiences into order. As such, personal constructs represent attempts to develop as internal ideas an understanding of the world around them. A technique that was used in connection with the principles of PCP is that of the 'repertory grid'. This relies on the use of bipolar constructs (e.g. hot and cold) to understand the internal workings of each person's personal construct. The approach used in the study was to identify particular points in time in the move process. By the use of 'snap cards', people were asked to consider several specific points in time. Six points in time were identified including their previous workplace and an alternative workplace that they may have chosen to work at in situations where the new workplace was unsuitable or less convenient. The point in time considered included:

Previous workplace → Anticipated workplace → Workplace on move day → Workplace after several weeks → Workplace after several months → Alternative workplace

Respondents were asked to envisage how they felt about their particular work environment at different points in the move process. Each of the participants in the study were asked the following question:

Which two of these workplace environments (work settings) are the same in some way and different from the third?

This question was repeated for various combinations of triad to allow a full exploration of the constructs associated with their feelings towards the workplace and changes to them over time. One example of the resulting repertory grid for a single respondent is given in Figures 5.4 and 5.5. The figures show four

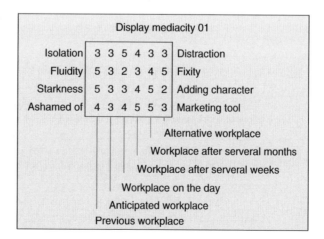

Figure 5.4   Repertory grid showing changes in attitude towards the new workplace.

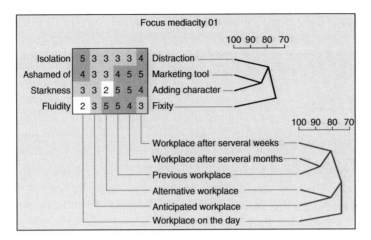

Figure 5.5   Repertory grid showing clustering of constructs relating to the workplace move.

specific bipolar constructs (e.g. fluid versus fixed). Each of the points in time (e.g. workplace after several months) is then rated on a value scale from 1 to 5. A score of 5 indicates that the word on the right (e.g. fixity) most accurately describes the characteristic of the space at the particular point in time. Conversely, a score of 1 in this case suggests that the leftmost construct (e.g. fluidity) most accurately reflects the prevailing characteristic of the workplace at the particular point in time.

Similar repertory grids were obtained for twelve subjects in the study. Several recurring themes emerge when looking at the constructs construed by the twelve respondents in relation to the movie experience:

- Organisational pressures that overlay the move process in a climate of uncertainty.
- Ambivalence towards new work environments that introduce leading-edge technology whilst necessitating conformance to a less personalised setting.
- A sense of loss of ownership in relation to 'time' as well as 'space' necessitated by the free-address system.
- Acceptance that the use of high value assets (multi-media studios and equipment) necessitate a level of sharing to enable leveraging of asset values.
- Evidence of a sense of loss, with those affected by the break-up and reassembly of work teams across a split site.

To date, the particular research tool has provided insight into the transition process. Whilst the issues identified above are typical of some of the emerging issues, a continued longitudinal analysis will be required to understand the transition journey of individuals affected by the relocation. A similar study by Pepper (2008) illustrated how employees struggle to reconcile their own experiences of a new facility in relation to five problematic themes: transition, openness, finances/layoffs, management and the building itself. One key finding was that users of the building experience the facility very differently from the way that the designer had originally intended.

5.7 Conclusions

In this chapter we have considered the central idea of the 'service concept'. This concept is presented as the driving force behind any planned change in facilities management. As such, it provides a proactive starting point. A multi-stage model is described, which can be used to assess the robustness of such a service concept, allowing further refinement where necessary.

Exploring the effects of any change initiative, we are forced to consider the impact on stakeholders and the core business. Invariably this impact is felt through the physical space, which itself is the subject of change. This change process continues well beyond the relocation or move stage. Rather than being a one-shot process, evidence suggests that it is an ongoing evolutionary process. Space is contested, modified, usurped, experienced and attached to. The day-to-day influence of the facilities manager and users alike gradually eclipse the intentions of the original facility designer.

Understanding the temporal nature of space becomes key to unlocking its potential. For some organisations the pursuit of novelty and experience justifies the continued 'scenery changes'. An alternative opposing facilities management strategy might seek to reinforce attachments: attachments that provide a familiar 'homely' setting. As to which of these two diametrically opposed approaches prevails depends upon broad-ranging organisational intentions. Does place attachment foster organisational loyalty? Is organisational loyalty desirable or does it discourage the free transfer of human assets? What is clear from our consideration of space and physical assets is that such assets can determine the success or failure of an organisation's change in the management process. Through an appreciation of the relationship between users and space we can ensure that such changes result in the desired outcomes.

References

Adams, J., Hayes, J. & Hopson, B. (1976) *Transitions: Understanding and Managing Personal Change*. Martin Robertson, London.

Allen, T. J. (1977) *Managing the Flow of Technology: Technology Transfer and the Dissemination of Technological Information within the R&D Organization*. Research supported by the National Science Foundation. MIT Press, Cambridge, MA. 329 pp.

Altman, I. and Low, S. (eds) (1992) *Place Attachment*. Plenum, New York.

Armenakis, A.A. and Harris, S.G. (2002) Crafting a change message to create transformational readiness. *Journal of Organizational Change Management*, **15**(2), 169–183.

Becker, F.D. (1981) *Workspace: Creating Environments in Organizations*. Praeger, New York.

Bitner, M.J. (1992) Servicescapes: the impact of physical surroundings on customers and employees. *The Journal of Marketing*, 57–71.

Brown, G., Lawrence, T.B. and Robinson, S.L. (2005) Territoriality in organizations. *The Academy of Management Review*, **30**(3), 577–594.

Clark, G., Johnston, R. and Shulver, M. (2000) Exploiting the service concept for service design and development. In *New Service Development: Creating Memorable Experiences* (eds J. Fitzsimmons and M.J. Fitzsimmons). Sage Publications Inc., Thousand Oaks, CA.

Duffy, F. (1997) *The New Office*. Conrad Octopus, London.

Elrod, P.D. and Tippett, D.D. (2002) The 'death valley' of change. *Journal of Organizational Change Management*, **15**(3), 273–291.

Foucault, M. (1980) *Power/Knowledge*. Harvester Wheatsheaf, London.

Friday, S. (2003) *Organization Development for Facility Managers: Leading Your Team to Success*. Amacom Books.

Gibson J. (1979) *The Ecological Approach to Visual Perception*. Houghton Mifflin, Boston, MA.

Hall, E.T. (1963) A system for the notation of proxemic behavior. *American Anthropologist*, **65**(5), 1003–1026.

Hammond, G.D., Gresch, E.B. and Vitale, D.C. (2011) Homegrown process improvement employing a change message model. *Journal of Organizational Change Management*, **24**(4), 487–510.

Hatch, M.J. (1987) Physical barriers, task characteristics, and interaction activity in research and development firms. *Administrative Science Quarterly*, **32**(3), 387.

Holm, I. (2006) Ideas and Beliefs in Architecture and Industrial Design: How Attitudes, Orientations, and Underlying Assumptions Shape the Built Environment. Oslo School of Architecture and Design.

Jay, M. (1986) In the empire of the gaze: Foucault and the denigration of vision in twentieth-century French thought. Foucault: *A Critical Reader*, 175–204.

Jeffreys, J.S. (1995) Coping with Workplace Change Dealing with Loss and Grief. Crisp Learning.

Kelly, G.A. (1955) *The Psychology of Personal Constructs*. Norton, New York.

Kotler, P. (1997) *Marketing Management: Analysis, Planning, Implementation, and Control*, 8th edn. Prentice-Hall, Englewood Cliffs, NJ.

Kubler-Ross, E., Wessler, S. and Avioli, L.V. (1972) On Death and Dying. *Journal of the American Medical Association*, **221**(2), 174–179.

Lefebvre, H. (1991) *The Production of Space*. Wiley-Blackwell.

Markus, T.A. (1993) *Buildings and Power: Freedom and Control in the Origin of Modern Building Types*. Routledge, London.

Milligan, M.J. (2003) Loss of site: organizational site moves as organizational deaths. *International Journal of Sociology and Social Policy*, **23**(6/7), 115–152.

Nevis, E.C. (1987) *Organisational Consulting: A Gestalt Approach*. Gardner Press, London.

Pepper, G.L. (2008) The physical organization as equivocal message. *Journal of Applied Communication Research*, **36**(3), 318–338.

Perlow, L. (1998) Boundary control: the social ordering of work and family time in a high-tech corporation. *Administrative Science Quarterly*, **43**, 328–357.

Preiser, W.F., Rabinowitz, H.Z. and White, E.T. (1988). *Post-occupancy Evaluation*. Van Nostrand Reinhold, New York.

Rose, N. (1990) *Governing the Soul*. Routledge, London.

Stuart, F.I. (1998) The influence of organizational culture and internal politics on new service design and introduction. *International Journal of Service Industry Management*, **9**(5), 469–485.

Sundstrom, E., Burt, R.E. and Kamp, D. (1980) Privacy at work: architectural correlates of job satisfaction and job performance. *Academy of Management Journal*, 101–117.

Tannenbaum, R. and Schmidt, W.H. (1973) How to choose a leadership pattern, Institute of Industrial Relations. *Harvard Business Review*, May–June, 162–164.

Taylor, S. and Spicer, A. (2007) Time for space: a narrative review of research on organizational spaces. *International Journal of Management Reviews*, **9**(4), 325–346.

Toffler, A. (1970) *Future Shock*. Amereon Ltd, New York.

Vischer, J.C. (2008) Towards a user-centred theory of the built environment. *Building Research and Information*, **36**(3), 231–240.

Zeisel, J. (1984) *Inquiry by Design: Tools for Environment–Behavior Research*. Cambridge University Press.

Zeithaml, V.A., Bitner, M.J. and Gremler Dwayne, D. (1996) *Services Marketing*, International Editions. The McGraw-Hill Companies.

6 Managing Briefing in Major Projects

6.1 Introduction

6.1.1 Aims

Facilities managers should be able to:

- ensure the client and other key stakeholders, such as users, play a full and productive part in the briefing process;
- successfully manage the dynamics of the briefing process and the associated team building required;
- ensure that the necessary information is collected and presented in a way that enables the client to visualise what is proposed.

6.1.2 Context

As Chapters 2 and 3 have shown, it is important, for effective facilities management, to understand and respond to the needs of the relevant stakeholders. This is the case in general, but becomes particularly pointed when a major step change is planned in building provision, be that a new completely new building or significant alterations to an existing building. In this situation it is likely that the consultation will intensify, with the possibility of productive engagement, but also the risk of raised expectations and dashed hopes. Briefing is the *process* by which the stakeholders' requirements are translated into a tangible, built solution. Figure 6.1 indicates a fairly typical journey for clients (Barrett and Stanley, 1999), which resonates with Figure 5.4 in the last chapter.

It can be seen that initially expectations rise as all the possible options are considered and then it progressively becomes apparent that not so many of the benefits that seemed to be on offer will actually materialise. Further, the practical disruption and uncertainty induced by the changes slowly become apparent and before long the whole process is looking quite irritating for the users. This

Facilities Management: The Dynamics of Excellence, Third Edition. Peter Barrett and Edward Finch.
© 2014 John Wiley & Sons, Ltd. Published 2014 by John Wiley & Sons, Ltd.

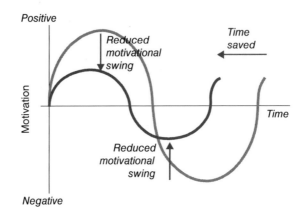

Figure 6.1 Motivational swings in the briefing journey.

can make it harder for those driving the project, as the relationships become strained, leading to problems in communications and understanding, which in themselves demand attention to repair. Gradually the project is realised and strangely most people involved are quite pleased with the outcome, but are scarred by the journey they have been through. Indeed, in a project looking at theatre projects a client suggested that it was as if they had been 'mugged' by the construction industry (Short, Barrett and Fair, 2011). In another case a client was heard to say that, although the project had ultimately been successful, they felt like 'paupers in a palace', owing to hard-to-find running costs.

It should be possible for the briefing process to be handled with more finesse. This chapter therefore aims to show how the benefits of reduced motivational swings and a faster, more efficient process can be achieved, eating up less organisational goodwill and steering the project to an improved outcome that really fits with the organisation's imperatives. This is suggested by the dashed line on Figure 6.1.

6.1.3   Summary of the different sections

Briefing will be defined below and five action areas described that are key to successful briefing (including the above issue of effectively handling the progressive dynamics of the process). The section will finish with some practical briefing tools. Case study examples are drawn from various organisations and settings.

6.2   The importance of briefing

All organisations will at some time find it necessary to prepare a design brief. This may be because the company wants new, specially designed premises or wishes to upgrade its existing accommodation in some way. The design of buildings is a contributing factor that can affect an organisation's ability to achieve its goals. Therefore, when new building work is being considered, it follows that if an organisation wishes to obtain value for money and a building to meet its requirements, then time and effort should be put into the briefing process.

At its most basic level, a brief is essentially a statement setting out the client's requirements for a new building/refurbishment/room layout. However, this is a very limited view of the subject and can lead to a negative use of the brief as a way of controlling the client. A broader, more productive approach results if briefing is seen as a process, defined as follows:

> Briefing is the process running throughout the construction project by which means the requirements of the client and other relevant stakeholders are progressively captured, developed and translated into effect.

This describes a creative process from which it is hoped good ideas will emerge as the pooled abilities and knowledge bases of the client, facilities manager, design team, etc., interact over time to find solutions that improve the infrastructural support provided to the client. The end result should be satisfied stakeholders, not only in terms of the end result, but also the process. What could be more critical than this?

Given his deep knowledge of the organisation and its needs, the facilities manager has a very important part to play in the briefing process. Depending on the definition of the facilities manager's job, this could amount to providing information about, say, the user's input or it could involve running the whole process. In what follows it has been assumed that the facilities manager has scope to influence all aspects of briefing.

6.3   Five keys to successful briefing

Every organisation has its own particular approach to briefing and success can be achieved in many different ways. However, it has been found through in-depth research that there are five common areas through which improvement can be sought from whatever base. These are shown in Figure 6.2 and are described in detail in *Better Construction Briefing* (Barrett and Stanley, 1999).

6.3.1   *Empowering clients*

Built environment professionals have a habit of blaming clients for not being experienced. This is like saying people should not buy cars unless they are competent mechanics. It is for the professionals to do all they can to help the client perform their role, building from the fact that as the client they will know their *own* business but not the built implications of their ideas.

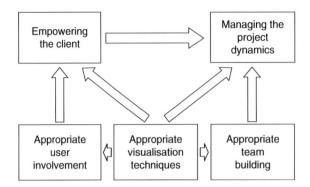

Figure 6.2   Key improvement areas in briefing.

Of course a client's lack of experience may cause problems. However, several studies have been undertaken where communication patterns between clients and professionals were analysed during the briefing process (Newman *et al.*, 1981; Gameson, 1991; Farbstein, 1993). These studies found that the input from the parties varied considerably depending upon the client's previous experience of construction. In cases where the client had no previous knowledge or dealings with the building industry, the consultant (architect, quantity surveyor, structural engineer, etc.) tended to dominate any discussions. The opposite was found to be the case where the clients had previous experience. Therefore, it is important to establish at the outset what level of experience the client possesses. If a consultant is dealing with an inexperienced client it is important that a common language is established between participants to avoid misunderstandings.

The potential benefit of increasing communications between the different parties is shown in Figure 6.3 by the Johari Window (Bejder, 1991). The public area represents the client communicating his initial requirements without difficulty to the design team. The blind area can be seen to be the needs of the client that are identified by the consultant (architect, engineer, etc.) through two-way discussion. The private area relates to information that is not disclosed by the client, whether intentionally or not, until a good level of trust has developed. The unknown area is initially hidden from both parties, but may reveal itself through joint discussion once a good working relationship has been established. The more information a consultant has about a client the better the final brief is likely to be, as it will be tailor-made for that specific organisation.

To be *effective* clients should:

- be knowledgeable about their own organisation;
- be aware of project constraints;
- understand their roles and responsibilities;
- maintain participation in the project;
- gain the support of senior managers;

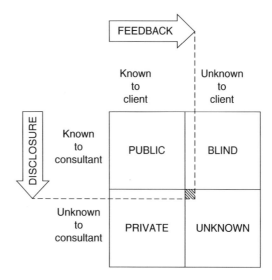

Figure 6.3   Johari Window as used in the briefing process.

- appoint an internal project manager;
- integrate business strategy and building requirements.

Clearly, facilities managers can have a significant impact on several of these fronts. Much of the knowledge about the organisation and its needs is available at first hand and facilities managers' typically technical background should make them an effective conduit for ensuring that the project constraints and roles and responsibilities are fully taken on board by the client. Facilities managers are in a good position to ensure senior management support is obtained and maintained, especially where they themselves hold senior positions in the organisation. Indeed, facilities managers may actually carry out the internal project management role and can be a focus for stability in the relationship between the project-focused participants and the permanent staff of the organisation. Facilities managers should be the natural focus for integrating building requirements with the strategy of the client organisation.

Thus, there is very great potential for the facilities manager to positively empower the client either as a consultant or as a key member of the client organisation. However, implicit in the need for an active and informed client/ facilities manager is the need to *invest* time and energy in their contribution to the project in hand. This may require planning to create the slack for someone relatively senior and very familiar with the organisation to have time allocated for them to build up their knowledge of the project and a shared understanding with the design team as to what is desired. A good example of this is given in the case below, which goes on to illustrate other aspects of good practice as well.

Case Study: The Chinese Arts Centre (CAC) in Manchester (Barrett, 2004)

The CAC opened in 2003 and provides a new, innovative, multi-functional home for a national charity established in 1986 to develop and promote Chinese arts in the UK. The scheme comprised in part the conversion of the ground floor and basement of a Victorian building fronting the street (see Figure 6.4a), and also the fitting out of commercial space to the rear, which was part of a new mixed-use development. The work took place in 2002 and 2003 after well over a year of searching for a suitable location and developing the ideas and design of the scheme.

The scheme 'jigsaws' various spaces together cleanly into a compact scheme that makes use of all available space. Access and security are well thought through: the shop/tea area/gallery are all overlooked by the offices through horizontal slot windows (see Figure 6.4b), meaning that 'invigilators'

(a)

(b)

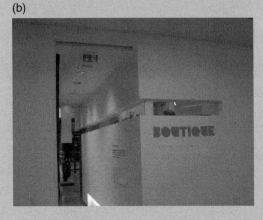

Figure 6.4   Photo (a) and (b) of the Chinese Arts Centre, Manchester (14 and 21).

are not needed in the gallery, which the CAC could not have afforded. Rear access to the gallery is provided, meaning that the open space to the back of the CAC could be used for performances. The spaces are generally simple and passive in decoration, but include strong elements of colour, some traditional Chinese references and lively lighting using lanterns.

From the start, the project was led internally by a 'Project Champion', the Chief Executive Officer, Sarah Champion, who the Board freed up to investigate and research the scheme. Early on, the architectural practice OMI were appointed as advisors. A secondee from a housing association also provided guidance on general issues relating to the construction industry. In this way the client's resources were made available, and appropriately augmented, right from the earliest stage. The Project Champion literally walked the streets searching for potential properties to house the organisation. Multiple assessments were made, in close interaction with OMI, allowing the needs of the organisation to be clarified and the nature of the scheme that would satisfy these needs established. In formal terms, a full Development Appraisal was produced to set out this thinking.

Impressively, the core design team and the Project Champion were funded to spend a week in Beijing. This experience proved very valuable for all involved. The architects gained a first-hand sense of Chinese design whilst interacting with the client representative in a variety of settings, so capturing a rich understanding of her aspirations. The trip had a number of tangible impacts on the design in, for example, such features as grey backdrops picking up the brick and stone colour in Beijing, punctuated by blood red accents. Similarly, the floor pattern was based on the brick layouts seen in courtyards, while artefacts from Chinese buildings were reused in the scheme, notably in the doors to the basement education rooms.

As the preferred option for the location of the new CAC was finalised, a range of other consultations were carried out. This included visits of the CAC Board to other galleries to test out their preferences, plus a visit by the designers to Zumtobel Lighting's Austrian headquarters to review 'daylight' lighting systems for the large galleries. There was also a series of consultations with specific stakeholders on a needs basis throughout the design development. As OMI or the client's Champion identified issues that needed clarification, a group would be formed and the particular issue discussed, clarified and then resolved. This happened many times, with tangible outcomes. For example, a group of potential artists-in-residence were brought together to work through the provision they would need to live in the gallery. This discussion resulted in a highly innovative solution that has proved very popular with the artists.

The construction contract was let on a traditional basis. Owing to the fairly prolonged gestation of the project, the design was well developed and most of the details had been worked through. A contingency of 5% was explicitly included and some other figures allowed for extra flexibility in the light of unforeseen issues. A good relationship existed between the contractor and the architects and there was no project manager. Given this degree of preparation and the strength and compactness of the team of people involved, the construction phase ran very smoothly, albeit there was, of course, the need to constantly respond to relatively minor issues and to ensure that they were resolved within the clear cost plan for the project. In the event the project was completed on time and £196,000 under budget. The CAC's Champion successfully negotiated for the release of this margin by the ACE and it was used for community art around the building.

The shortlisting of the Centre for the RIBA Inclusive Design Award describes the gallery as 'a real *tour de force*. It is a beautiful space with a high level of built-in spatial and service flexibility, all with the feel of the Orient'. The CAC has been well received by its staff, artists and clients. The facilities have proved flexible and have been intensively used. The design won an RIBA Award for Design Excellence in 2004.

The client carries out surveys of users twice a year and the design and atmosphere of the building is consistently positively received. The scheme was limited by the constraints of the site and the existing buildings, but, by working closely with the client, OMI have successfully accommodated the organisation's needs whilst creating a set of integrated and characterful spaces. Similarly, the artists-in-residence are very positive about the innovative, self-contained, living and working solution.

The background research revealed that there is not a particular tradition of gallery-visiting in China. In addition, there had been problems of security in CAC's previous accommodation, but the Centre could not afford invigilators. The design solution of viewing slots (long horizontal windows) between the central offices and the various spaces responded to both these issues, successfully drawing the public into and around the building, and providing security in an economical way.

The large gallery is big enough to be used in a variety of formats for shows of different scales. Private views are a good test and the space successfully accommodates between 50 and 400 people as the visitors ebb and flow. In practical terms, the technicians find the wooden wall surfaces robust and flexible for taking down and creating new displays.

The running costs and maintenance of the building have been straightforward, although the underfloor heating is thought to be slightly inflexible and expensive. It was provided as a result of the client's insistence that the building be easy to operate given that there is no resident caretaker. The client's view is that 'the building has worked well' and has successfully adapted to changes in demand.

The design process was notable for the active and continuous presence of a highly committed client representative, who brought an intimate understanding of the client's business and also was willing to learn about construction. She put in considerable research and sheer legwork to provide the designers with a clear brief, albeit one that evolved iteratively as a result of close interaction between the client and designers over an extended period and with the added input of a wide range of targeted user consultations. Given this preparatory work the design was well developed by the time the contract was let. The construction phase went extremely well owing to this clarity of specification and the existence of a clear cost plan, against which any variations could be carefully managed, utilising the reasonable contingency as appropriate.

In the example given in this case study the client's representative was the Chief Executive in quite a small organisation. In larger organisations it would be inconceivable that the CEO could take up this role, but as already mentioned the organisation's facilities managers could have many of the attributes that would enable them to effectively carry out this bridging role between the organisation and the built environment professionals engaged on the project. Facilities managers should, for example, have a close understanding of the organisations' operational imperatives and a working knowledge of the technical aspects of the built environment. Of course to fully carry out the role of the professional client the facilities manager also needs to have a full understanding of the core business of the organisation and its strategic intentions.

Clients are not always expert in the area of construction and the built environment, but they are pivotal to a successful briefing process. To play their role fully, clients need to be empowered, and facilities managers have an important part in this, either through consultancy advice or by acting as the 'expert client' of the organisation for which they work.

6.3.2 *Appropriate user involvement*

Empowering clients involves the provision of a focus within the client organisation where the various views of the stakeholders that make up the client system can be synthesised. When considering the building process, there are normally three different factions involved: designers, paying clients and end users. Traditionally, as shown in Figure 6.5, there has been very little communication between the end users and the other two groups (Zeisel, 1984). The designers and paying clients have made the users' decisions for them without proper consultation. Hence, the final users have often found that the new building does not meet their true needs. This has resulted in costly alterations after a project has been completed. Such problems lead to the conclusion that end users should be involved in the briefing process. Many studies have been carried out on design projects where users have participated in the briefing process. It has generally been concluded that users were happier with buildings where they had been involved in decision making (Farbstein, 1993).

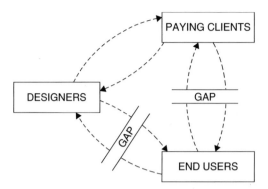

Figure 6.5 The user–needs gap.

Thus, a major stakeholder perspective that must be kept in focus is that of the user. This section focuses on users, but the same principles are likely to apply to other stakeholders. It can be thought that more is better when it comes to user involvement. This is not so. It is quite easy to get together groups of users and involve them in consultation where the usual office status hierarchy skews the discussion. Disappointment sets in as it becomes clear that, in practice, involvement does not equate with influence. This is the situation shown in Figure 6.1 with the unhelpful motivational swings. This is an easy trap to fall into, as simply by asking someone what they would like the impression can be created that they are likely to get their wishes. The solution is to manage expectations very carefully. In one project studied the following simple formula was used in the consultation with users: 'We probably won't be able to provide it, but in an ideal world what would you really want?' This revealed a lot of wishes (often clashing), but opened up the possibility of responding to many of these in the design, at least to a degree, all within an atmosphere of realism. This was felt to be far better than either of the alternative solutions of: not asking or only surfacing mundane ideas from the users.

Appropriate user involvement calls for:

- understanding the benefits of the investment;
- an assessment relative to each situation or issue;
- careful handling of group dynamics;
- maintaining user involvement throughout the project.

The benefits of user involvement should be the starting point and these come in two broad forms. Firstly, there is a technical dimension where technical information about the needs and wants of the users can be obtained and, secondly, there is the social or political dimension of developing an acceptable solution. To achieve the first, technical, aspect argues for specific groups of people brought together with particular knowledge to address a clearly focused issue. For an open discussion it is important to aim for groups that are of relatively equal power. To achieve a politically acceptable solution, active feedback to the groups

Case Study: Using customer feedback to tune the retail environment (Barrett, 2007)

Sainsburys have around 35000–40000 customers passing through their doors and operate in an extremely competitive environment. For them property is a very significant investment, but is clearly seen as a means to the end of selling to consumers. It is therefore instructive to analyse how Sainsbury's approaches this challenge. The fundamental basis of the approach is to constantly listen to their customers (and staff) and understand their needs and then to exploit the continuous flow of project work on their stores to very efficiently meet those needs. The investment of time and energy to capture those needs is extensive:

- For a *new* store:
  - Market researchers will interview a 200 potential customers making up the likely profile for the location.
  - Those 200 will be brought in for facilitated focus groups in groups of 30–50. the fixed parameters are spelt out: e.g. overall size of shop and then through discussion issues such as the following are revealed:
    - The retail offering for which there is demand in the context of local tastes and competition – will a fish counter be popular; is there a popular local butcher already?
    - The local employment situation for staff.
    - The local traffic issues – when is congestion high re schools/ factories and will bus routes work or even be relevant? What sort of parking provision is favoured?
    - Important aspects of the environment of the shop, layout, aisle widths, lighting, etc.
    - Local preferences about produce reflecting regionality.
  - Similar discussions are held with staff.
- For a *refurbishment* the same process is followed, but the discussions with existing customers and staff are much more focused as they have direct experience of the store and a range of practical improvement possibilities are revealed.
- For each *existing* store about ten calls to customers are made each week informed by the company's record of their shopping behaviour to understand better the impact of the store on their propensity to buy. This is reinforced by specific data on sales every time a change is made anywhere in a store, say undershelf lighting on to high-end toiletries, which led to higher sales.
- Each year two *concept* stores are developed to trial new ideas and for these the consultation and monitoring is especially intense, including full-scale mock-ups in a warehouse around which feedback is sought from 200 customers and staff.

● Finally, and perhaps the most impressive indication of commitment to understanding customers (and retail staff), all 2000 staff involved in corporate services from the CEO down, and including property of course, spend every Friday of every week in the stores working alongside the staff on the ground.

Given all this information the question is how to use it to inform the design and operation of the stores. Here a decision has been taken to create a stable construction flow so that the learning can articulate with it in a progressive manner. Therefore stores are updated at five years and fully refurbished every ten years. With around 460 stores this generates an annual programme of work of around 35 light and 65 intensive refurbishments, plus around 30 extensions and 15 new stores. Blueprints have been created for each generic size of store and these are developed and finessed so that the delivery of the stores is as efficient as possible. At Tesco's this led to a reduction in build cost from £250 to £115 per square foot over five years. Taking a holistic view of engineering services, rather than letting consultants address lighting, heating and ventilation and cooling separately, led to savings of £1.5 m per store. Beyond this, framework arrangements led to very competitive pricing by suppliers, based on bulk purchasing, and with chosen consultants by providing a reliable flow of around 80% of the work. For standard items, however, competition is opened up to additional vetted bidders and e-tendering on an auction basis is increasingly used.

Have the customer's views really made any difference? The extensive consultation described above allows stores to be tuned to local circumstances, but a lot of the feedback does confirm what is already known; however, there is 'something always something new'. Out of these processes the evidence has been gathered to: change the tradition of providing 'warm' lighting to an emphasis on 'natural' lighting; redesign layouts so that sweets are not placed next to checkouts as it irritated adults with children at a complicated part of the process of shopping; (from staff feedback) change layouts to keep stockrooms close to frozen goods to minimise warming time; progressively introducing self-scan checkouts as customers like them (even though it takes longer!); trialled popular concept by building a store around the 'worlds' within one store to give a local specialist shop feel, but found that in practice efficiency was more important to customers – the same applied to more complex aisle layouts.

Sainsburys appear to invest heavily in both listening to customers and trying things out experimentally in a restless search for better built solutions to meet customer needs. In many ways customers want simplicity and stability in their shopping experience so that they can efficiently and quickly get the shopping done. Thus designing stores that work smoothly and intuitively is a major aim and feedback from customers is key to understanding and achieving this. There is a belief that stores can be 'better (for customers) – simpler (for the business) – cheaper. If we are obsessed with the customer then we can usually get all three.'

about the impact of their contribution is needed. This needs to be maintained throughout the project so that the transition from initial discussion to actually moving in is smoothed. As a little example, it could make sense for the technical designers working on finishes to meet with a group of cleaners to specifically discuss the floor/wall junction. The following case study gives a more extensive example drawn from the dynamic world of retailing, where gaining potential customers' views can be problematic, but vital to success. The scale of the effort here may be beyond many organisations and (as will be seen) links into a programme of development, but the shape of their multi-faceted, layered, purposeful approach is instructive for all scales of consultation.

This case study stresses a belief in the multiple benefits that can be realised through a clear focus on a major overarching goal. This issue of focus will infuse the next section on managing the project dynamics.

6.3.3 *Managing the project dynamics*

A recent study of capital arts projects (Short, Barrett and Fair, 2011) worked through their files and tracked the budget for each project throughout its life. It can be seen from that for Poole 'Lighthouse' Arts Centre in Figure 6.6 that these can be incredibly volatile. On this project the initial construction budget was around £10 m, but early on this dropped to just over £5 m; then as the project progressed it

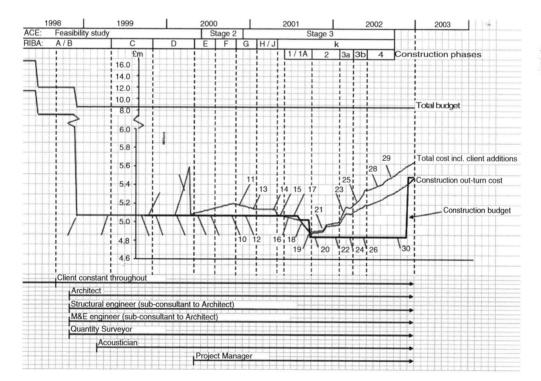

Figure 6.6   Cost profile of the Poole 'Lighthouse' Arts Centre Project.

spiked at £5.6 m, but was brought back through aggressive cost control, the budget then fell to around £4.9 m before rising steadily in the construction phase to nearly £5.7 m. This was actually a very well managed project (as will become apparent below), but even so the dynamic nature of the process is clear to see.

More than that the study of these projects clearly identified that trying to fix the brief more and more definitely at the earliest stage is in fact counterproductive. Time and again the pattern in Figure 6.6 was seen, of spikes in budgets aggressively brought down and then costs steadily rising back to a similar level. In most of these cases the cost control was carried out under the rubric of value management, but was primarily cost cutting. In the process many benefits were lost and interactions between aspects of the design challenged, e.g. the acoustic isolation needed between activities. As time went on many of the savings had to be reversed and the costs would rise again, but the coherence of the design could prove difficult to recreate. So how could we say that in the case of Poole it was well managed?

The answer is that, in the case of Poole, the client had made a similar investment to Manchester Arts Centre with the technical director provided with extra time and support to focus on the project. In addition, the force of circumstances created a very extensive early consultation phase that was scaled to a project around twice as big as the one that was actually realised. Building on this, the client created a robust matrix of functional requirements against which changes for cost or other reasons were judged so that any decisions respected a coherent prioritisation of the client's vision. This was linked to an undeclared contingency sum (not known to the design team), which, together with any savings, allowed the controlled release of resources towards the end of the project against a list of priority items that were reintroduced as the certainty of completion approached.

What was special about Poole's priority matrix? It did two things that traditional construction costing by element, or area, do not. These were, firstly, to be driven by a 'weighted matrix of competing stakeholder needs and aspirations that accommodated economic, political, and practical views' and, secondly, these were resolved to single weightings around 106 subprojects that each related to a distinctive functional possibility. This could be the creation of a theatre space, or something like the upgrading of façade as a whole. This exercise entailed detailed manipulation of the costing information by the quantity surveyors. Decisions were then taken balancing these possibilities and the benefits they would bring to the successful operation of the theatre, all within the available budget. As a result, when decisions were taken the coherence of the whole operation or its design were not lost. Also big decisions could be taken with confidence, such as that to leave the appearance of the façade pretty much as it was. As the Technical Director said: 'We get our money for what we do inside the building, not what it looks like outside.' A version of the priority matrix is given in Table 6.1.

This example from Poole is not a pro forma document for others to use, but it can be seen as a pro forma process of thinking that others can learn from in order to successfully navigate the, often, stormy passage through their project. In more general terms it is crucial to see briefing as a *process* not a document or an event. Seen like this the dynamics of the process becomes very important and should

Table 6.1 Priority matrix for the Poole 'Lighthouse' Arts Centre.

	Cost £	Economical	Political	Practical	Weighting
1. Flexible performance space	99,000–1,140,000				
Shell	600,000	6	9	8	1
Link	50,000	5	3	8	1
Dressing rooms	100,000	3	1	4	3
Timber infill to stairs	5,000	0	0	4	1
M&E	inc. in shell	1	2	8	1
Seating, stage lighting, fitting out	50,000–200,000	6	3	9	5+
Landscaping	25,000	0	0	0	7
Theatre technical systems	160,000	7	2	9	1
2. Wessex Hall	893,000				
Staging	25,000	6	4	6	5
Seating	250,000	10	10	10	1+
Acoustic enhancement	58,000	3	10	8	1
Disabled access/Balcony works	25,000	0	5	0	5
Theatre technical systems	485,000	3	0	9	1
General redecoration	50,000	9	9	9	1
3. Wessex backstage	252,000–353,000				
Instrument store relocation	60,000–85,000	0	0	6	5
Toilets	12,000–18,000	6	10	10	2
Dressing rooms and general refurbishment	180,000–250,000	2	4	8	1
4. Towngate Theatre	650,000				
Acoustic enhancement	35,000	6	7	3	1
Disabled access	20,000	10	10	10	1
Theatre technical systems Seating	570,000	3	0	9	1
General redecoration	25,000	9	9	9	1
5. Towngate backstage	180,000–350,000				
New Lift (to plant room level)	100,000	0	3	8	1
Disabled access and facilities	30,000–50,000	1	8	8	1
Upgrade showers and dressing rooms	50,000–200,000	5	2	6	2
General redecoration		9	9	9	1
6. Façade	530,000–715,000				
New lifts	200,000	4	5	5	1
Staircases	125,000	2	8	5	3
Toilets	110,000	2	8	5	3
Brise-soleil	10,000–30,000	2	8	5	3
New cladding inc. structural frame	Not ab option	–	–	–	–
Maspny cladding	15,000–20,000	2	8	5	3
Window repairs/replacement	20,000–120,000	2	8	5	3
New entrace doors	40,000–100,000	2	8	5	3
New signage	10,000 up	2	8	5	3

Key

1+ Vital to our continued business.

1 Of great imporatance to PACT.

2 Important that these areas are refurbished but not necessarily requiring a rebuild.

3 Whilst acknowledging the importance of the façade as a whole these items are dependent on other parts of the scheme.

4 Of vital importance but there may be a more economical way of providing them.

5+ Could manage without if absolutely necessary.

6 Would be important, if dividing screens happen.

>7 Come bottom or our list of requirements.

run in parallel with the whole construction project, although of course the most intense activity will be towards the beginning and the process must include progressive fixity. What is required is a live record of the consensus, the shared vision held by the client, other stakeholders and the design team. This can then drive the design process, without constraining developments in the ideas.

To approach the dynamics in this way requires teams to give various aspects attention, namely:

- establishing major project constraints early on, including critical dates;
- agreeing procedures and methods of working;
- allowing adequate time to assess client's needs;
- careful validation of information with the client organisation;
- feedback to all parties throughout the project.

Facilities managers are well placed to establish major constraints, such as key dates, the overall budget and the space and flexibility available. It is almost unbelievable, but projects have hit major problems over confusion as to whether VAT was included in the budget or because a key university committee was missed and a year was lost! A clear protocol for interacting throughout the project is rarely created, but is not difficult and this should allow adequate time at the early stages for the client's needs to be assessed. There is always pressure to move quickly, but time spent early on to allow ideas to grow is time well spent on all but the simplest projects. Facilities managers can be in a position to argue for this space to be provided.

Communications must be interactive with extra efforts made to ensure that the client's apparent statements are clearly understood. This should link to additional efforts to make sure all are kept informed through feedback. The temptation is to short-circuit these efforts over and above simply getting an answer from the client and acting on it, but this will only lead to conflict and dissatisfaction later. Again facilities managers should be advocates for this investment, which may not already make immediate sense to those in client organisations.

6.3.4 *Appropriate team building*

As buildings increase in complexity, it becomes less likely that a single person could possess all the knowledge required to design a new building. Therefore, most projects will require a team of designers if all the different requirements in a client group are to be satisfied. As design teams increase in size, communication within the design team becomes as important as communication between the designer and the client. The following groups of people may need to be consulted during the briefing process:

- Architects
- Project managers
- Quantity surveyors

- Structural engineers
- Town planners
- Mechanical engineers
- Building surveyors
- Electrical engineers
- Interior designers
- Acoustic engineers
- Landscape architects

The choice of a design team is critical to the success of a project. Clients should take time to select an appropriate team and try to achieve a balance between technical knowledge and team skills.

> The prime objectives of clients should be to understand their own needs first and then to secure a design team who will reflect their view of the world. Clients who choose a team at odds with their own world view run the risk of being swept away by the design team's own enthusiasm down inappropriate directions. The client has a real choice and he should use it (Powell, 1991).

Many client organisations emphasise a hands-off, lowest-price tendering approach to amassing the project team. This can result in superficial contractual arrangements between the organisations and a shifting mesh of participants as these organisations move staff from project to project to balance their internal demands. The result is that although the arrangements may look tight on paper, in practice the individuals involved keep building relationships and shared understanding and then breaking up so that the tacit understanding within the project team constantly erodes. The result is not always optimal! Ideally:

- selection should focus on assembling complementary skills, not simply on lowest price;
- team members should approve and understand the management structure of the project and should, as far as possible, remain constant throughout the project.

These are areas where facilities managers can collect data on project participants to inform careful selection. They can ensure that project team members are clearly briefed as to the role of their input to the project. They can also strongly advocate that the client demands key individuals who will remain consistently involved.

As an example, in one project studied the client organisation had, over the years, used a number of different building consultants. As hospitals are such specialised buildings, it was corporate policy to reuse good consultants as they would already be familiar with the difficulties of healthcare design. In theory this seemed sensible, but in practice it created problems. As the consultants were normally only recalled every few years, the specific individuals who had worked on previous projects had often moved to different consulting firms. This meant

Case Study: Multiple players and multiple issues

The importance of creating a strong team, made up of individuals with the capacity to effectively work together and bring with them complementary knowledge and skills, is brought into sharp focus by the project histories set out in Figure 6.7 (Sutrisna and Barrett, 2007). These highlight the fact that as well as all the project issues around technical and economic matters, there is a key overarching role of creating, maintaining or enhancing the core business, or in the case of the Contact Theatre, having to close down the business and reinvent it, not helped by a fire halfway through the project! In all this cycling between various issues, the current version of the 'vision' is where pressures are resolved – or not.

Following the storylines 'C' and 'L' presented in Figure 6.7, at first glance it is immediately apparent that both projects have gone through complex processes. The two construction projects were of the same building types and were funded with a significant proportion of public funds but differing in financial values. The construction project of the second case study (delivered using the management contracting route) cost roughly ten times the first case (delivered using traditional procurement) in terms of financial values. The drivers ('1' in Figure C) of the first case include the declining income (due to local competition of similar and alternative 'products' and also a low subsidy from the government through an agency), the poor (existing) condition of the building facilities and a threat from the building owner to use one of their building facilities for other purposes in the near future. The drivers of the second case came purely from the local city council who were inspired by various successes of similar venues ('2' in Figure L) to catalyse their ongoing regeneration ('3' in Figure L) and promote the local heritage at the same time. Prior to the availability of the public fund for their specific type of buildings, European funding (ERDF; '4' in Figures C and L) seems to have been the originally intended source of funding for both projects, which turned out to be a disappointment in the first case, but became an important funding element of the second case project. The selection of the architects brought distinctive features to the projects. In the first case, the selected architect had experiences in designing for sustainability (involving a natural ventilation system; '5' in Figure C), whilst in the second case project, the local heritage ('6' in Figure L) added another dimension and the selected architect is well known for 'signature architecture' ('7' in Figure L).

From the comparison of the two resulting figures, it is clear that the client in the second case project was experienced and tried to empower themselves by recruiting various experts – with a relatively better governance structure ('8' in Figure L) – thus surrounding themselves with various consultants. In contrast with the first case, the client relied on the architect and a smaller group of consultants ('9' in Figure C) to provide solutions. The availability of the public money for that specific type of building ('10' in Figures C and L) had ignited both projects to carry on with their plans. Whilst the further development of both projects had resulted in enlargement of both projects at some points, in the first case this also involves increased complexity ('11' in Figure C), resulting from the adaptation of the natural ventilation system in the design to achieve high visual impact (architectural merit) to reinforce their application, but within a tight timetable (rushing – '12' in Figure C – to be among the first applicants to increase their chance to secure the funding). The tight

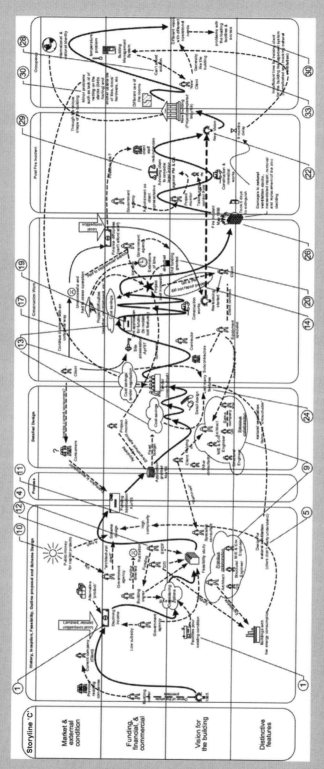

Figure 6.7 Rich pictures of the Contact Theatre (Sutrisna and Barrett, 2007).

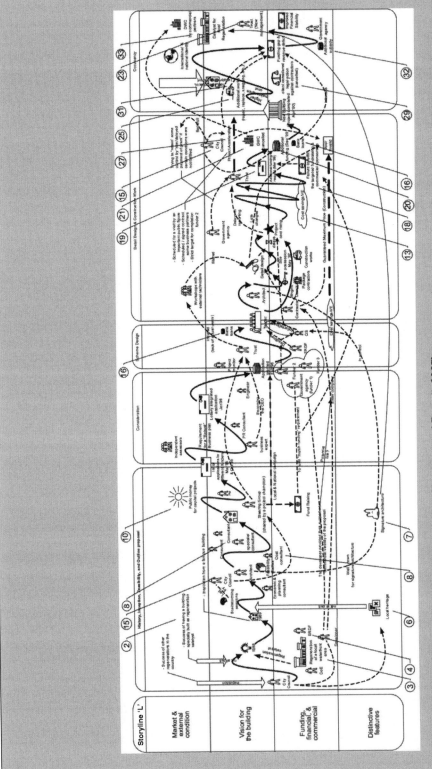

Figure 6.7 *(continued)*: Rich picture of The Lowry (Sutrisna and Barrett, 2007)

timetable, deadlines and complexity of both projects contributed to the delays in issuing details and underestimation by the quantity surveyors. This eventually led to costs rising and the necessity for cost savings exercises ('13' in Figures C and L) as the budget was 'fixed' from the funders. It is also noticeable that there were 'delays' in reporting the financial condition to the clients of both projects, which may have led to a false sense of security. In the first case, the client could not even understand why the cost savings were continuously implemented as a result of their lack of experience, whilst the project manager and the quantity surveyor failed to provide 'client-friendly' reporting ('14' in Figure C). The client in the second case was in a better bargaining position towards the funders due to the support from the city council ('15' in Figure L), including in financial matters, even though they needed to take on some bank loan ('16' in Figure L) facilities at some points.

It is evident in both case studies that the construction projects had contributed to the financial difficulties for the client organisations for different reasons. The client of the first case project failed to generate enough income from the continuation of their business in an 'unfamiliar' way ('17' in Figure C) as their buildings were closed for construction. The client of the second case project failed to achieve the planned fundraising and commercial sponsorships ('18' in Figure L). Clients of both projects had to come back to the funders for additional funding ('19' in Figures C and L), which was then absorbed ('20' in Figures C and L) by the construction process to complete the projects. This indicates the insufficiency of the earlier construction cost estimates on both projects and the flexibility of the funders over their original 'fixed budget' policies. In the second case project, one of the funders (funder 2) even allowed the diversion of the budget ('21' in Figure L, originally to construct a peripheral building) to the construction of the main building as a part of the additional funding. In the first case project, the construction project even contributed to the transformation of the client organisation into a new organisation with a new vision ('22' in Figure C). During the occupancy period, the client of the second case project experienced a change in their governance and management arrangements ('23' in Figure L) as an indirect result of the funding gap and deficit in the project. A combination of cost savings/value engineering (VE) and redesign was implemented in both case projects to deal with the rising costs. In the first case project, the client ended up losing many of the original features ('24' in Figure C), while in the second case project, the client managed to put back most of these ('25' in Figure L) at a later phase. The delays in the first case project, which were aggravated by the fire incident ('26' in Figure C), proved to be detrimental for the client organisation's financial condition, leading to mass redundancies of their staff. In the second case project, the client initiated acceleration ('27' in Figure L) to achieve the targeted completion date, contributing to the later financial deficit of the organisation. The indirect effect of the delays in the first case project involved temperature problems resulting from insufficient time to conduct proper training of the client about how to use the complicated building management system. The only solution to this problem, which is to recommission the entire system, has been found too expensive ('28' in Figure C) for the client organisation. It is also noticeable that both clients were considering legal actions ('29' in Figures C and L) at some points. Even though both legal actions were cancelled (as proving negligence may not be a straightforward task), they indicate the adversarial nature of the process in both cases despite the different procurement routes taken.

The transformation of the client organisation in the first case project has resulted in different types of operations with different needs and created some problems ('30' in Figure C), such as difficult loading facilities and access, different use of the rooms and also some minor problems due to the earlier cost savings that have contributed to increased day-to-day maintenance costs. In the second case project, various works had to be performed ('31' in Figure L) after practical completion to address the extensive snagging list, repair works and replacement works necessary to achieve the required standards. These additional works eventually contributed to the clients' financial difficulties that necessitated them receiving a significant additional sum from the government agency (funder 1) to improve their financial stability ('32' in Figure L).

However, despite the various problems encountered during the construction processes and after practical completion, both clients expressed their satisfaction ('33' in Figures C and L) with the resultant buildings. The new/refurbished buildings have been considered excellent tools to create/promote their international and national identity.

that the appointed consultants had to go through the learning process again. The organisation relied on the fact that the consulting firms possessed the necessary knowledge, when really it was only the individual.

6.3.5 *Using appropriate visualisation techniques*

A key blocker to successful briefing is the failure of the communications used by the parties involved. This is particularly acute in the area of visualisation, for the simple reason that the briefing process is a journey from a set of ideas in various people's heads to a single physical reality – a building or new space. Representing these ideas in terms of written specifications, two-dimensional drawings, etc., is not very reliable. Different people gain different impressions from each of these mediums. For example, those in the construction industry are used to 'reading' drawings and assume that others can too, but in fact many people cannot imagine three-dimensional spaces from a flat drawing. They may never have been through that journey from plan to place before. Even within the construction industry silly things happen, like colour-coded services drawings copied in black and white!

Getting to a shared consensus through these media is not easy or reliable. The consequence is that very often the built artefact that results is not what many of the participants expected, leading to confusion and disappointment or, less often, to a pleasant surprise. Unfortunately it is, more often than not, the client, maybe inexperienced in construction, who is least able to foresee what is going to result until it is too late. This can lead to problems of late redesign and severe inefficiencies.

Table 6.2 (Short, Barrett and Fair, 2011) draws a parallel with the process of creating a play as set down by Stanislavski (1936). The challenge is analogous to briefing, in that something excellent, but initially ill-defined, is jointly sought and a balance

Table 6.2   Stanislavski and Briefing. Drawing from the classic text for actors An Actor Prepares (Stanislavski, 1936), the following table draws parallels between Stanislavski's advice to actors engaged in the process of creating a play and the approach that should be taken by those involved in the creative act that is a capital project, especially in the early stages.

Element of Stanislavski's system	Parallel in briefing for capital arts
Preparing by exploring the 'given circumstances' of the play and experimenting with them creatively through purposeful action (p. 104) using the 'magic if' (pp. 46–47; 51–52)	Building the client's experience and capacity in the pre-design phase, by allowing time to both reveal the context factors through wide stakeholder consultation and to experiment with alternatives for various aspects
Progressively building up of an 'unbroken line of circumstances' linked to a 'solid line of inner visions' to illustrate each part (pp. 63–64; 68)	Begin brokering a vision in pre-design by surfacing the volatile/multiple visions of the stakeholders involved. Maintaining involvement of key players and especially the client
Using circles of attention ('nearest', 'small', 'medium', 'large') – progressively broadening the scope included, but returning to a smaller focus if attention begins to waver, and then building again. Removing the physical tenseness that 'paralyses our actions', especially at times of great stress (pp. 81–84; 96–99)	Cope with complexity by using visualisation tools to help clients and other stakeholders understand the progressively widening connections between the parts. Employ staged design management that allows those involved the time and conducive environment so they can take stock calmly and objectively
Choosing major units, each with creative objectives, that 'mark your channel' towards the main objective. Using only as many as necessary to avoid losing a sense of the whole/core of the play. Using verbs to ensure objectives are 'lively'. Linking parts by 'blocking out a line of physical actions'. Keeping to essentials – 'cut 90%' (pp. 114–117; 123; 142; 162)	Provide careful project management that employs 'value engineering' at intervals to ensure that the functional parts of the design are optimally contributing to the overall aims of the project and that any extraneous elements are pruned. Ensure that these exercises add up around the vision into an authentic 'value management' process
Creating an 'unbroken line of communication' on the stage that ebbs and flows continuously and exhibits real communion/understanding between the players, who adapt to the variety of relationships that confront them (pp. 196; 201–203; 224; 254)	Maintain the synchronisation of the inputs of multiple stakeholders. To do this, actively avoid superficial consultation and, as far as possible, keep the individual players and their shared tacit understandings constant throughout the project
Through an arduous joint search, a shared 'super-objective' and 'through line of action' should emerge. These then provide a strong, unifying focus for all of the subsidiary 'units', their objectives and associated activities. They also actively engage the individual trajectories of the actors, with the thrust of the play (pp. 271–273; 301; 306–307)	Settling on a shared vision after extensive experimentation and interaction of the key stakeholders and then guarding this vision, but also allowing it to evolve as appropriate, throughout the design and construction phases into use. Infuse the design with wide user consultation without moving away from the established vision

has to be maintained between the detail and the coherence of the whole design. What is clear is that continuous rounds of interaction using different tools to support visualisation are needed. As with briefing as a whole, it is a dynamic process.

Thus, the appropriate and timely use of a range of visualisation techniques is a major aspect of communications that, if successfully addressed, can make a huge difference to all other aspects of the briefing process. Again the suggestion is not that more communication is better per se, but rather that appropriate techniques should be used. In facilities management this analysis has some simple implications:

- basic techniques should be used carefully;
- extra efforts should be made to help inexperienced construction participants to visualise what is being proposed.

Many of the normal techniques used, such as drawings and specifications, can be very useful, but need to be used with great care and sensitivity, taking into account the experience of the participants. This may call for simply explaining to people some of the construction conventions used or orientating people so they know which way up the plan is, where the ground and first floors are and so on. Beyond this, simple techniques that are not very commonly used can make a huge difference. For example, showing a variety of photographic images, taping out the size of the new office space on the floor or, more elaborately, mocking up an office space with some furniture or visiting a similar building. This last can be very powerful, especially if those involved make the visit/s together and can discuss alternatives (compare this with the Chinese Arts Centre case above).

The effort should ideally be continued throughout the project, including site visits (potentially raising health and safety issues) so that the users can continue to input to the more detailed design decisions towards the end of the project and also so that they can 'close in' on what the end result is going to be like and gain a sense of ownership.

Case Study: Visualisation for users

This project was a specialist theatre conversion of a mill building where a wide range of consultation techniques was used with staff and potential users of the new accommodation, namely:

- Initially a scale model was built and this helped those involved to understand better the relationships between the proposed spaces.
- Then, when the space became available, the proposed rooms were taped out within the building and were mocked up with timber struts and paper. This stage provided the client with great comfort that the scheme was going to work and also allowed changes, e.g. the introduction of some curved partitions.
- As design neared completion the architects provided three-dimensional virtual fly-throughs, which supported discussion of such issues as colour schemes.
- Lastly, site visits were made to the facility (especially designed for disabled users), who on visiting late in the project pointed out that the light switches were being located too high on the wall for wheelchair users to reach – fortunately it was not too late to correct this.

Facilities managers can have a key role in arguing for the importance of spending time and other resources in order to make sure that those who are not so experienced in construction are given the chance really to know what is being proposed when they are asked to make serious, important decisions.

6.4   Management of the briefing process

Briefing is not a rigid process. Each person who deals with briefs will use a slightly different approach or emphasis. Also every project is different, hence any briefing model will require alteration to suit a particular situation. However, there are some basic guidelines that should be considered during the briefing process. The following examples can be used to demonstrate these important points:

- traditional briefing process;
- phased briefing process.

6.4.1   Traditional briefing process

The RIBA Plan of Work, shown in Figure 6.8, outlines the building procedure and associated briefing process that is generally followed in the UK (other countries adopt similar procedures) (Zeisel, 1984). It can be seen that the main brief is developed through the four initial stages: inception, feasibility, outline design and scheme design. A breakdown of what is included in each of these stages can be seen in Table 6.3 (Salisbury, 1990).

As a process model, the RIBA Plan of Work is helpful to a client or facilities manager, as it demonstrates the various steps that they should go through. However, this method is not actually very realistic. It does not make it clear enough that the client will be required to make frequent evaluations during the process. It mentions amendments and appraisals, but it does not suggest that the client may be forced to rethink some of their requirements quite late on in the briefing process. It is not always possible to achieve a perfect match between the design of a building and the brief requirements. Conflicts may occur between space requirements or the budget may not allow for all briefing requirements to be included.

It should also be taken into consideration that the actual process of collecting briefing information may raise users' awareness of their situation. For example, while considering their requirements, clients/users may begin to question how logical their working practices actually are and decide to change how they operate. Therefore, if a brief is fixed early on, as it is in this example, major changes in work practices can be difficult to achieve. In using this method, clients are encouraged to obtain as much information as they can early in the process; consequently, the brief can become too detailed too soon. Overdetailed briefs sometimes preclude creative solutions that were not obvious early on in design development. The next section aims to address the above problems.

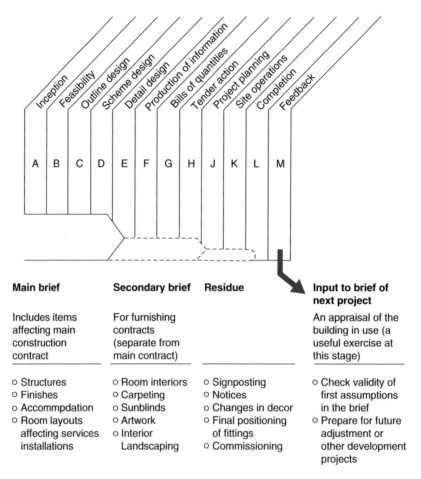

Main brief

Includes items
affecting main
construction
contract

○ Structures
○ Finishes
○ Accommpdation
○ Room layouts
   affecting services
   installations

Secondary brief

For furnishing
contracts
(separate from
main contract)

○ Room interiors
○ Carpeting
○ Sunblinds
○ Artwork
○ Interior
   Landscaping

Residue

○ Signposting
○ Notices
○ Changes in decor
○ Final positioning
   of fittings
○ Commissioning

**Input to brief of
next project**

An appraisal of the
building in use (a
useful exercise at
this stage)

○ Check validity of
   first assumptions
   in the brief
○ Prepare for future
   adjustment or
   other development
   projects

Figure 6.8   Development of the brief with regard to the RIBA Plan of Work.

6.4.2 *Phased briefing process*

This method follows similar steps to the traditional method, but the importance of evaluation is made explicit. The phased method was developed by the Stichting Bouwresearch (Building Research Board) in Rotterdam, as the result of a research project into client briefing (Spekkink and Smits, 1993). The basic principle of the system is that the brief contains, prior to each new planning phase, only the minimum amount of information necessary to be able to direct the plan in the next phase. The project is therefore considered in very general terms during the initial phase, working towards greater detail with every respective phase (see Figure 6.9). For example, if the subject of location/spatial relations was being examined, the following issues would be considered in each phase:

- Phase 1: the relationship between the proposed site and the rest of the town;
- Phase 2: the relationship between the proposed site and the immediate neighbourhood;

Table 6.3   Development of the brief through RIBA initial stages A–D.

Client action	Material for brief	Consultants action
Stage A – Inception		
• Considers need to build • Sets up supporting organisation (working party, committee or representative) • Appoints consultants • Commences exchange with consultants • Provides information for outline brief	• History of events leading to decision to build • Details of client and consultant firms, personnel • Timescale for the project *Outline brief* • Policy decisions • Purpose and function of project • Details of site and services • Basic details of building requirements, and lost limit	• Carry out preliminary consultations and appraisals of buildings • Receive and examine outline brief
Stage B – Feasibility		
• Conducts user studies • Considers feasibility results and analytical studies and reports • Develops brief	• Additions/amendments to outline brief in as much detail as possible about: site conditions; space requirements; relationships and activities; interior environment; operational factors • More precise information about client's financial arrangements	• Survey and study site and locality • Consult statutory authorities • Conduct feasibility exercises and studies of features of the brief • Advise about meeting of cost time limits • Elicit information required, and guide and assist with collection of briefing material
Stage C – Outline proposals		
• Receives and appraises designs and reports • Receives and approves outline designs and costs	• Amendments and additions to brief as a result of appraisals • Completed room data sheets	• Produce first sketch designs for analysis • Complete outline design and cost plan • Complete informal negotiations with statutory authorities
Stage D – Scheme design		
• Recevies and approves full scheme designs and costs (if satisfactory) • Instructs preparation of presentation drawings • Authorieses formal submission for required statutory consents	• Amendments and more details • Layouts etc. of furniture and equipment in special rooms and areas	• Prepare full scheme designs and estimate of costs • If approved, prepare presentation drawings, perspective sketches and /or models • Apply for planning consents

After the completion of Stage D correlation between the brief and the scheme design should be complete.

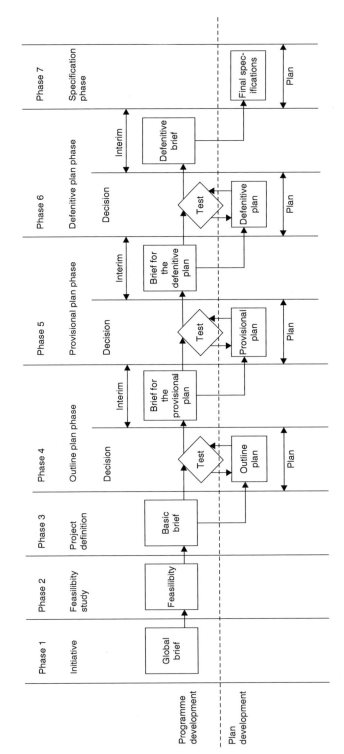

Figure 6.9   Phased briefing process.

- Phase 3: the relationships between different departments;
- Phase 4: the relationships between people or spaces within a department;
- Phase 5: the relationships between pieces of furniture in a specific workspace.

Such a phased development results in a number of advantages for clients and users. A series of phases allows clients and users to become familiar with problems and to become aware of new possibilities that new or refurbished buildings could provide for them, including changes to their own organisation. In addition, this means that users and clients can introduce their requirements at strategic moments, rather than having to consider all their options from the outset.

It is imperative that the process is tightly controlled: a situation whereby clients introduce requirements at the wrong time must be avoided. Once major planning decisions have been made they should stay fixed. Process control makes it necessary to ask the right questions at the right time. Hence, this briefing technique can be seen as a process of two-way communication between parties. Phasing offers clients and users the opportunity to assess the new building several times before the design is firmly fixed, allowing adjustments to be made if necessary.

The facility manager should implement an iterative staged briefing programme, where the brief is assessed frequently to ensure that it still matches the organisation's requirements.

6.5   Information required during the briefing process

Once a suitable process for briefing has been selected it is necessary to establish what information needs to be collected. Obviously every building project is different. An organisation may only want to refurbish existing offices or else it may wish to commission a new prestigious headquarters. In both cases it is necessary to collect a certain amount of data in order to be able to prepare a good brief so that the organisation gets the exact design it wanted. This section provides a guide to the information that facilities managers should collect during the briefing process.

There are four distinct areas where information may need to be collected:

- organisational concerns;
- external influences;
- individuals and work styles;
- physical environment.

Case Study: Management of the briefing process

The following description outlines the approach to briefing of the hospital organisation used for case study examples in Chapter 3. In order to remain competitive and meet healthcare legislation, this organisation is almost constantly involved with updating/altering its building stock. This means that the organisation is very familiar with the practice of briefing and hence over the years has developed its own particular briefing method. Its approach is used in the case studies to highlight good briefing procedure as well as the occasional mistake. The different stages are detailed below and relate to Figure 6.10.

Normally the hospital manager will identify the need (1) to improve a hospital and will contact the regional facilities manager to discuss what work is necessary. Their proposals will then be put to the corporate facilities manager who will appoint a project manager (2). The latter works with the hospital manager to produce a feasibility study, which demonstrates the financial implications (3). Once the feasibility study has been approved the project manager works with the hospital manager and the matron to produce a general policy statement, which describes which parts of the hospital are to be altered and why (4).

Discussions then take place between the project manager and the relevant departmental heads to obtain an overview of each department and a policy statement is completed, which describes how the department will be run. The project manager then produces an outline brief, which is tested at a series of meetings to ensure that the overall approach has been agreed (5).

Departmental heads are then asked to complete room schedules for their department, stating what rooms are required and what each room will contain. The project manager tests these requirements against a set of documents called the Health Building Notes, which contain preferred sizes for specific rooms, as well as comparing them to plans of existing hospitals. All of this information is then fed into the provisional brief, from which the project manager produces a provisional plan, which shows the location and sizes of all rooms (6).

At this stage department heads will often pin up relevant parts of the provisional brief on the notice board so that they can obtain staff views on the proposals. The brief is amended as necessary, resulting in a definitive brief. It is only at this stage that the architect is brought in to produce presentation drawings. The quantity surveyor is also engaged to advise on costs (7). The definitive brief and scheme are then reviewed at regional and corporate level. If they are approved the scheme is worked up in detail in preparation for tender (8).

Comment

The organisation has developed an approach over the years that allows it to build up the brief in stages, ensuring that the views of all interested

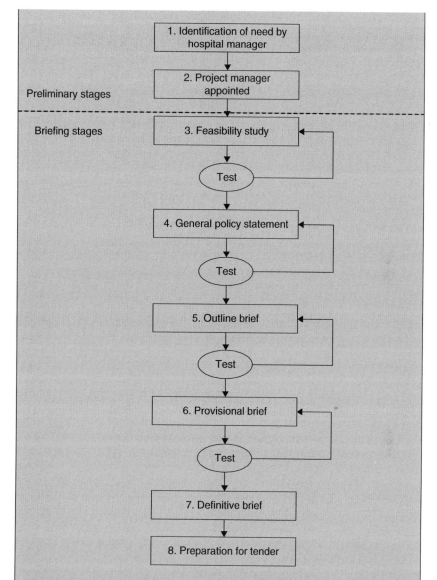

Figure 6.10   Briefing method as discussed in the Case Study.

parties are taken into consideration. By following this method the project manager is not overloaded with all of the information at once. People are asked for their opinions at the most appropriate stage; only the hospital manager and department heads are allowed to comment on general policy, while nursing staff are asked for their opinion on the workings of specific rooms. This process is very similar to the phased briefing process described earlier.

Table 6.4 Information to be collected during briefing.

Organisational concerns

- *Future plans/objectives* – What are the organisation's plans for the future? Are there any initiatives already in place that the general workforce is not aware of? Does management intend to alter the structure of the organisation, resulting in a reduction of the workforce? If so, how does this affect space requirements?
- *Fixed constraints* – What decisions are fixed before briefing even starts? What finances are available for the project? What is the intended time span of the project? How many people does the building need to accommodate? Will the old furniture be utilised? Establishing what things are fixed will enable facilities managers to direct their data collection efforts towards areas where future decisions will need to be made.
- *Corporate culture* – Every organisation will have a different culture, which affects how the organisation operates. It will be reflected in the daily actions of management and staff: how do people interact with one another, who makes what kinds of decisions, how do people use their time/space? It provides clues during interviews, for example, whether peoples' opinions reflect those of others or whether they hold conflicting views. Data collection methods can also be influenced; interview data may be useful in one company, whereas quantitative survey data may be the norm in another.
- *Organisational structure* – The briefing team needs to know how an organisation is structured. What sort of decisions are taken at which level? Who has to report to whom? This kind of information can determine from whom information is collected and about what.
- *Staff* – Headcount projections for the next few years are of course a necessity. However, it is also helpful to know what sort of employees make up an organisation. Are the staff professional people who have very high workplace expectations?
- *Image expectations* – Organisations may wish to project a specific image to the outside world, and hence will require certain aesthetic qualities from their building.

External influences

- *Laws and codes* – It is necessary to be aware of any government legislation etc. that is likely to have an effect on the design of the building. This could include zoning plans, environmental legislation and fire regulations.
- *New technologies* - Changes are occurring in technology all the time; what is today's standard may have been superseded tomorrow. Organisations should seriously think about what is likely to happen over the next few years, so that the new building design can accommodate changes without major constructional alterations.
- *Labour force patterns* – Changes have been predicted in the labour force over the next decade. There are likely to be, for example, far more female workers, perhaps resulting in increased safety requirements within workplaces. Alternatively, an increase in older workers could lead to concerns about health, fitness, lighting and air temperature.
- *Competitors' actions and plans* – How does the organisation compare to its competitors? Are their facilities better equipped to do the job? If so, are they likely to attract our customers or employees? It is also useful to learn from competitors' experiences; has their new approach to space planning improved their service?

Individuals and work styles

- *Task analysis* – What exactly do different members of staff do? Specific job descriptions are useful for understanding how people fit into an organisation.
- *Environmental satisfaction* – How satisfied are staff with their current environment? Does their immediate environment enable them to work effectively and productively or have they just learnt to adapt to an environment that was poorly designed in the first place?
- *Communication and adjacency patterns* – Who communicates with whom, where, when and how often? Does the environment assist with communication or hinder it? Are the right departments/people situated together?
- *Space, furniture and equipment requirements* – what furniture and equipment is necessary in order for an individual to perform his job? Has enough space been allocated for specific tasks?

Table 6.4   *(cont'd)*.

Physical environment
• *As-built plans* – These are useful as a comparison tool. They can be used to test how proposed space requirements relate to existing ones. Can a department justify its requests for twice as much space as they occupy at present? • *Space standards* – Space requirements should be recorded for all proposed areas. Remember to include support spaces, such as conference rooms, break areas, etc. • *Furniture and equipment inventory* – Can any of the existing furniture be utilised in the proposed new building? Which pieces can be used in their present form? Which will require refurbishment? • *Amount, type and variety of IT* – This again requires an inventory to establish what exists at present. What is to be relocated? Do any pieces of equipment require special environmental conditions? • *Circulation requirements* – This applies to both people and objects. Do corridors have to be a certain width? Hospital corridors, for example, need to be wide enough to accommodate patient trolleys. What about ramps, staircases, lifts and hoists? • *Transportation and parking* – How often do deliveries take place? Is a special loading bay necessary? How many car parking spaces are required for staff? How many people use public transport instead? • *Surrounding amenities* – Information about the external environment will govern what services are provided on site. It is important to understand how the surrounding area may change in the near future. Is a public car park being planned? Are there so many local facilities for food that staff are unlikely to use a cafeteria. • *Appearance* – Factors relating to preferred form, scale, texture, colour, proportion and style of building, both internal and external.

Table 6.5   **Information to be considered by specialist consultants during briefing.**

Specialist considerations
• *Loading requirements* – superimposed loads, wind loads, acceptable limits of structural deformation • *Fire protection* – legislative requirements: means of escape, fire-fighting installations and equipment, spread of flame prevention • *Contamination protection* – ventilation, humidity, proofing against contamination and damage, security installations • *Heating/cooling* – thermal factors, temperature requirements • *Lighting* – natural and artificial lighting requirements, emergency lighting, display lighting • *Acoustics* – factors affecting sound attenuation, sound insulation • *Energy demands* – services required, standby systems, solar energy requirements • *Maintenance needs* – maintenance programme demands, quality of structure, materials and installations

These are listed in more detail in Table 6.4. These lists are by no means exhaustive and should be regarded more as a prompt or checklist to ensure that nothing has been forgotten. It should be remembered that allowances should be made for future use, as well as present requirements. It should be pointed out that various processes exist for collecting data, but because the same methods are used in evaluation, the different examples are described in Chapter 3, Section 3.4.

The client should be able to collect the necessary information on most of the above topics, aided by consultants where necessary, but there are certain specialist areas where consultants will have to advise the client. These areas are covered in Table 6.5 below so that the client can ensure that they have been considered.

Case Study: Collection of information for briefing

In order to provide the best possible service, hospital managers constantly monitor their establishments to see where improvements could be made. Recently, in one particular case, the hospital manager was finding it increasingly difficult to staff the two operating theatres, as they were located one above the other and so had to be staffed separately. The facilities department was asked to investigate the problem and suggest alternative plans. During their investigation it was discovered that the local National Health Service hospital possessed a clean-air enclosure, a facility that was not provided in this particular hospital. Therefore, the facilities department proposed that this would be a necessary addition to any refurbishment work, in order for the organisation to remain competitive.

Comment
The above example demonstrates that it is essential to consider external factors as well as internal ones. If the organisation had failed to identify the need for a clean-air enclosure, then it is possible that consultants would have chosen to send their patients to the National Health Service hospital instead.

Summary

1. The facilities manager should collect information in the following four areas:
    (a) organisational issues;
    (b) external influences;
    (c) individuals and work styles;
    (d) physical environment.
2. The facilities manager should check that consultants have addressed all of the relevant specialist issues.
3. The facilities manager should ensure that the information collected relates to future as well as present use.

6.6 Conclusions

The above advice has tried to balance covering some detailed aspects whilst maintaining a broad sweep. Although there are many other sources of much more detailed advice, one useful source that is resonant with the approach recommended here is the ACE/CABE *Building Excellence in the Arts: A Guide for Clients* (Short *et al.*, 2009). This is written for a particular sector, but much of the material is of general practical use, extending to strongly emergent issues such as accessibility and sustainability.

At the other extreme it should be remembered that any project is a huge, typically rare, opportunity to make a difference. Thus, every effort should be made to balance risk mitigation with actions to achieve an *excellent* result. In this

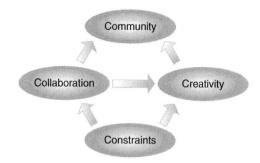

Figure 6.11   4Cs factors (Barrett and Barrett, 2006).

context it is instructive to finish with an analysis of the characteristics of exemplary projects – the very best projects in which nineteen very experienced practitioners have been involved (Barrett and Barrett, 2006). Four emergent themes are evident across these projects (see Figure 6.11):

- Constraints
  - ○ Generally resources and often time in particular, but sometimes issues such as physical limitations on the site. Often 'supernormal'.
- Collaboration
  - ○ Unusually close working together, typically involving the client, designers and contractors from early on in the project.
- Creativity
  - ○ Can be an innovative technical solution or procurement approach, but normally reflects a strong joint problem-solving approach throughout.
- Community
  - ○ Where the project positively impacts more broadly than the immediate participants in its long-term orientation and/or broader societal engagement.

At its richest, exemplary projects seem to be typified by significant constraints driving those involved to collaborate strongly, spurring the team to innovative responses that not only triumph against the demands of the project itself but also impact positively on the community around. This raises the interesting question in the project arena as to when a constraint is a positive thing and not a restraining force. The answer would seem to be when it is explicitly, clearly stated early on; when it is demanding enough to define the project, prioritise and reorientate behaviour around a superordinate goal and provide a clear measure of success; and when other (less important) parameters are dealt with flexibly to allow appropriate collaboration and creativity to meet the challenge. In this way the constraint has provided a clear space to work within and this certainty can clearly be stimulating provided sufficient flexibility with the remaining resources is available.

Clearly, achieving a successful – no an excellent – project outcome is not easy, but the results can provide enduring benefits and satisfaction that justify the investment demanded.

References

Barrett, P. (2004) *Chinese Arts Centre, Manchester: A Case Study*. CABE, London.

Barrett, P.S. (2007) *International Examples of Service-Driven Innovation in Construction*. NESTA, London.

Barrett, P.S. and Barrett, L.C. (2006) The 4Cs model of exemplary construction projects. *Engineering, Construction and Architectural Management*, 13(2), 201–215.

Barrett, P.S. and Stanley, C. (1999) *Better Construction Briefing*. Blackwell Science, Oxford.

Bejder, E. (1991) From client's brief to end use: the pursuit of quality. In *Practice Management: New Perspectives for the Construction Professional* (eds P. Barrett and A.R. Males). E.&F.N. Spon, London. pp. 193–203.

Farbstein, J. (1993) The impact of the client organization on the programming process. In *Facility Programming* (ed. W.F.E. Preiser). Van Nostrand Reinhold, New York. pp. 383–403.

Gameson, R. (1991) Clients and professionals: the interface. In *Practice Management. New Perspectives for the Construction Professional* (eds P. Barrett and A.R Males). E.&F.N. Spon, London. pp. 165–174.

Newman, R., Jenks, M., Bacon, V. and Dawson, S. (1981) *Brief Formulation and the Design of Buildings*. Oxford Brookes University, Oxford.

Powell, J. (1991) Clients, designers and contractors: the harmony of able design teams. In *Practice Management: New Perspectives for the Construction Professional* (eds P. Barrett and A.R. Males). E.&F.N. Spon, London. pp. 137–148.

Salisbury, F. (1990) *Architect's Handbook for Client Briefing*. Butterworth Architecture, London.

Short, A., Barrett, P., Fair, A. *et al.* (2009) *Building Excellence in the Arts: A Guide for Clients*. CABE, London.

Short, C.A., Barrett, P. and Fair, A. (2011) *Geometry and Atmosphere: Theatre Buildings from Vision to Reality*. Ashgate Publishing, Farnham.

Spekkink, D. and Smits, F.J. (1993) *The Client's Brief: More Than a Questionnaire*. Stichting Bouwresearch, Rotterdam, The Netherlands.

Stanislavski, C. (1936) *An Actor Prepares* (translated by Elizabeth Reynolds Hapgood). Methuen Drama, London.

Sutrisna, M. and Barrett, P. (2007) Applying rich picture diagrams to model case studies of construction projects. *Engineering, Construction and Architectural Management*, 14(2), 164–179.

Zeisel, J. (1984) *Enquiry by Design*. Cambridge Univetrsity Press, Cambridge.

Facilities Management Tools

7 Information Technology Tools for Facilities Management

7.1 Introduction

7.1.1 Scope of the chapter

The aim of this chapter is to explore the information needs of facilities management enterprises. In an increasingly diverse business environment it is often difficult to remain informed. Technical approaches often struggle to reflect the full breadth of issues and rates of change affecting facilities management. With this in mind, this chapter describes a 'holistic' approach to information technology governance when applied to facilities management. It details how a 'systems' approach might be used in practice. Such a system mimics natural systems rather than adhering to the mechanistic characteristics of current 'architectural systems', which have difficulty in representing how people share information in practice. After considering the information needs of facility users and information providers (data, information and knowledge) a systems model (viable system model) is described that offers a flexible and responsive approach in tune with a user-centric perspective. This IT governance model builds on the generic 'systems model' outlined in Chapter 1. Furthermore, it addresses some of the issues discussed in Chapter 5 (change management), which highlights the ambiguous nature of space as seen from different user points of view.

7.1.2 Summary of the different sections

- Section 7.1. Introduction.
- Section 7.2. A 'user-centric' approach to IT is proposed, identifying some of the pitfalls of conventional IT architecture when applied to FM.
- Section 7.3. Explores FM areas where IT has become an indispensable tool.
- Section 7.4. Highlights the emergence of two key concepts – interoperability and collaboration – driving development in FM information systems.
- Section 7.5. Examines the distinction between data and information, exploring how new IT applications in FM harness tacit knowledge.

Facilities Management: The Dynamics of Excellence, Third Edition. Peter Barrett and Edward Finch.
© 2014 John Wiley & Sons, Ltd. Published 2014 by John Wiley & Sons, Ltd.

- Section 7.6. Explains the principles of IT governance to ensure a user-oriented approach to IT development in FM.
- Section 7.7. Considers how the 'generic FM systems model' introduced in Chapter 1 can be applied to the FM IT context.
- Section 7.8. Conclusions.

7.2 Information philosophy

Information technology provides an indispensable backbone for facilities management decision making, enabling the sharing of information based on an armoury of technologies and communication channels. Multi-site portfolios of property are today overseen by occupiers and service providers alike, using complex enterprise-wide IT systems. Without the benefits of such systems, many of the advances in facilities management could not be fully realised. However, whilst IT managers within organisations may be evangelical about the promise of new technologies, those responsible for using the information and capturing it may be less enraptured by it. Fundamental to any successful facilities management information system is an understanding of how people use information rather than how they use an IT solution. The reluctance of users to accept information technologies is often seen by information managers as showing an 'irrational' fear. Such 'irrational' responses may in fact reflect very rational opposition to rigid structures and concerns about how people actually use information in practice. Davenport (1994) suggests that a 'user oriented' approach to information technology would help in overcoming many of the legitimate concerns shared by users of all organisations.

7.3 Modern applications of IT in facilities management

Today's facilities management information systems are a far cry from early computer-aided facilities management (CAFM) systems. With their arrival in the 1970s and 1980s, CAFM provided a much-needed capability of linking complex two-dimensional computer-aided design (CAD) floor layouts with databases. This allowed 'what if' scenarios to be developed regarding interior layout. For multi-storey office buildings this enabled careful crafting of building occupancy, based on what are known as 'stacking' and 'blocking' algorithms, which helped decision makers to optimally allocate space both vertically and horizontally within a building based on a detailed understanding of organisational requirements. Furthermore, such software enabled comprehensive and reliable inventories to be compiled regarding fixtures, fittings and cabling. For tenants, this allowed the tracking of assets; compilation of financial data for budgetary, tax and insurance purposes; as well as condition assessment for the purposes of health and safety and replacement strategies.

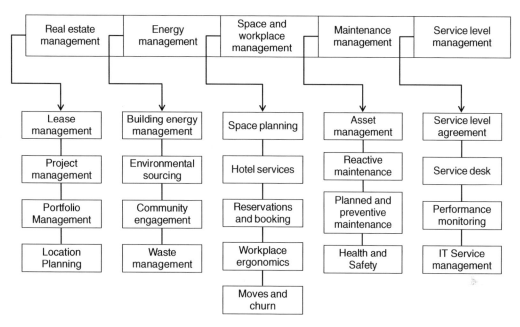

Figure 7.1   **Modules incorporated within modern computer-aided facilities management software solutions.**

Modern computer-aided facilities management now represents a major investment for owners of large property portfolios. The increased use of modularisation has enabled such software to meet the requirements of many different aspects of facilities management including:

- helpdesk administration;
- maintenance scheduling;
- administration of contracts;
- health and safety management and risk assessment;
- space management and room booking;
- procurement and management of stock;
- energy and environmental management;
- real estate portfolio management (including lease management).

Figure 7.1 illustrates the type of modules commonly used within a modern-day computer-aided facilities management solution. The modularised design of such systems allows exchange of information and the avoidance of data duplication between and within the modules. Going well beyond simple relational databases, these systems make use of an armoury of technologies. Field data can be collected using a combination of data logging/iPad devices used in conjunction with automated identification tagging systems (two-dimensional/three-dimensional barcodes, radio-frequency identification, near-field tracking). Similarly, detailed information and enquiries can be undertaken on site, by interrogating a remote CAFM system. Numerous Web enabled innovations also move CAFM from being a 'black box' technology such that building users can now engage in the facilities management process (e.g. kiosks, self-service facilities, multi-channel social networking feedback). Driving much of these efforts in IT innovation are:

- attempts by FM to get close to the customer (users and clients);
- attempts to increase the rigour and reliability of statutory reporting requirements (e.g. health and safety);
- attempts to harmonise information exchange (and performance standards) between service providers and clients;
- attempts to ensure information only needs to be entered once, thus minimising the costly and error-prone process of re-entering data.

Not only has the range of services covered within computer-aided facilities management increased in recent years but several other key developments characterise the modern-day landscape of facilities management information systems. These include:

- radically different approaches to the architecture of facilities management information systems, in particular building information management (BIM);
- extension well beyond simply software systems, where complex systems of data capture, data analysis and dissemination are now used as part of a complete 'system' allowing the management of information from its inception to eventual use;
- the evolution of many different forms of dissemination and communication (linked to the advent of social networking).

7.4 Interoperability and collaboration

For the most part, CAFM technology was deployed by office building occupiers at the fit out-stage. They sought to capture all relevant (and often irrelevant) data associated with occupying a new facility. In modern-day corporate environments pressures for globalised facilities management operations have displaced localised (building level) solutions that no longer address this much wider perspective. Common measurement standards and processes underpin a move towards enterprise-wide solutions that allow the management of a diverse portfolio, often spreading over multinational boundaries. The ability to benchmark and compare performance across a portfolio has given greater impetus to the creation of facilities management information systems.

Interoperability and collaborative working have become key words in an FM industry overwhelmed by change. Change orders have become a necessary feature of new builds, refurbishments and fit-outs, as organisations struggle to keep apace with turbulent business environments. As a consequence, designers, construction and facilities professionals end up working from different 'hymn sheets', inevitably giving rise to costly reworking and duplication of effort. It has become increasingly apparent that, given the pace of change, new more agile approaches to information management are required. Building information modelling (BIM) has been widely advocated as the solution to the problem of data sharing and collaborative working. A wealth of 'reliable and accurate' information previously 'locked' in drawings, documents and schedules can now be accessed by the facilities professional in a single digital model. The idea of reusing and re-purposing such data by means of a seamless transfer from the construction to the operation phase provides an enticing argument, giving rise to national initiatives to ensure this aspiration is realised. In a report by NIST (Gallaher and Chapman, 2004), the authors identified the potential for improvement by assessing the costs of inadequate interoperability in the US capital facilities industry. The headline figure revealed is $15.8bn 'lost' every year, or 1–2% of industry revenue. The report covers only commercial, institutional and industrial facilities, but the analysis extends to the whole life-cycle, from planning and design, through construction to operations, maintenance and decommissioning. A clear estimating methodology is given based on interviews and surveys. Interoperability is defined as 'the ability to manage and communicate electronic product and project data between collaborating firms and within individual companies' design, construction, maintenance and business process systems' (Gallaher and Chapman, 2004: p. ES-1). The estimates were arrived at by comparing the costs of existing practices with 'a hypothetical counterfactual scenario in which electronic data exchange, management and access are fluid and seamless'.

Barrett (2005: p. 32) points out that 'The results indicate that two thirds of the "unnecessary" costs are met by owners and operators of buildings, whereas architects and engineers are only associated with about a tenth of this impact. The balance of around a quarter of the costs is met by the industry in terms of contractors, specialists and suppliers.' The clear implication raised in the report is that 'interoperability costs during the O + M phase [result] as a failure to manage activities upstream in the design and construction process'. The figures also indicate that the motivation to move on the issue is skewed in that the greatest potential to change things is in the area where the least cost is experienced. However, interviewees expressed the view that, 'although a failure to exploit emerging technologies was an issue, it was a series of major "disconnects" in the processes that was the major disincentive'.

7.4.1 The principles of building information modelling (BIM)

Building information modelling describes an approach rather than a technology. However, it has only been possible with the advent of three key technologies:

- CAD
- Object CAD
- Parametric building modelling

Considering each of these in turn, we see how the approach has evolved from one based on two-dimensional electronic drawings of facilities to three-dimensional physical models and finally to 'intelligent' representations that are scalable and make use of intelligent objects.

BIM uses a modelling approach to manage and share information throughout the design, construction and operation phase. In essence, it is concerned with the management of information from 'cradle to grave'. The logic of such an approach is not only appealing to the construction industry but also to facilities management. Like the construction industry, FM is historically beset with problems of data loss, miscommunication and translation errors. Using the above technologies it is possible to minimise data re-entry and the errors associated with this. The underlying interoperability standards should, in principle, allow 'seamless' data transfer between phases. For the facilities manager, the enduring problem of coordination between components is addressed head-on. Changes during the design and construction process have often led to errors in coordination between building elements with component clashes and protracted snagging periods to resolve remaining problems. Such delays invariably impact on completion and move-in dates. Possession of a fully coordinated building model provides the basis for a facilities manager to then maintain all of the information relevant to the facilities.

7.4.2   Construction operations building information exchange (COBie)

A wealth of information resides in drawings, bills of quantities and specifications. Undoubtedly, this type of information is potentially useful to the facilities manager. However, this documentation is rarely compiled in such a way that it can be handed over to the client following the construction phase. In order to address this problem and ensure that such information remains useful, a singular method for collecting such information is essential. This approach sits at the heart of the UK's construction strategy (National BIM Report; NBS, 2012), which states that:

> … in order to improve the measurement and management of public assets, it is recommended that public clients request that specific information be delivered by the supply chain. The specified information set, called COBie, delivers consistent and structured asset information useful to the owner operator for post-occupancy decision-making.

The UK government has embraced the concept of BIM (as detailed in the Building Information Modelling (BIM) Working Party Strategy document), requiring three-dimensional BIM (including all project and asset information, documentation and data) for all public assets by 2016. At the heart of this strategy is a building information model known as COBie, the Construction Operations Building Information Exchange. One of the appealing features of COBie is that it is possible to represent the COBie model using a simple spreadsheet (e.g. Excel, Lotus Notes). In principle, this means that the possibilities afforded by building information modelling are available to even the smallest facilities management organisation. Following handover, the facilities manager is in receipt of information such as manufacturer's contact information, model and serial numbers of installed products and components identified by tags or barcodes. The facilities management team then has to append additional information to the model in the form of:

- updated space function and room area measurements;
- warranty, parts and hand-over documents;
- operation and maintenance documentation;
- installed product data.

Whilst COBie may not provide a 'silver bullet' for the facilities manager's information needs, it undoubtedly paves the way for improved sharing and collaboration with the design and construction stages. Many challenges remain, linked to the 'user oriented' demands of facilities management, but there is mounting recognition that BIM requires an accompanying cultural change. As observed by David Philip, Head of BIM Implementation at the UK Cabinet Office:

> One BIM input can give as many valuable outputs. In order for this process to be effectively implemented, however, it needs to be undertaken in a truly collaborative environment (with iterative feedback loops), and here lies the real challenge. Manifesting BIM beyond the technology and process to a cultural paradigm shift is where the real challenge lies. BIM is very much more a verb than a noun.

7.4.3 Remaining challenges in building information modelling (BIM)

Can building information modelling provide all of the answers to the information needs of a facilities manager? Is it simply a question of creating a vast and complex digital representation of the facility? Many advocates of BIM would suggest that the acronym actually refers to 'building information management' and that the intention is not to create a behemoth. Rather, at the heart of the concept is the ability to interoperate between systems, enabling a unified representation of the physical system. That being the case, BIM only provides the foundations or 'rules' by which information is exchanged in the design, construction and facilities management industries. The 'knowledge base' that underpins such models have yet to be fully established and BIM standards provide only the starting point.

This leads us to the role of the facilities manager. Will the advent of BIM displace their expertise? If anything, it will amplify their role as they seek to fill the void in knowledge and research that underlies the rules of a building information model (which is in fact not a single model but a range of physical, scheduling, cost, surface and behavioural models). Perhaps a more searching question is whether building information modelling provides the necessary flexibility to represent the complex nature of facilities management, being concerned with not only place, but also process and people.

Conflict 1: Shared meaning. If we confine our understanding of 'space' to 'space as distance', as described in Chapter 5, then facilities management activities can be fully represented in terms of a BIM model. Our attention is then drawn entirely to the physical characteristics and the physical condition of the built facility. However, as discussed in Chapter 5, 'space' as a concept is interpreted very differently by the various stakeholders in facilities management. We considered the idea of space as a 'controlling mechanism', as something that affects us 'emotionally' as an experience and something that lends itself to our immediate needs (affordances) irrespective of the original intended use of the space. Each of these differing perspectives gives rise to fundamentally different information requirements and interpretations. As observed by Davenport (1994: p. 122): 'No matter how simple or basic the unit of information may seem, there can be valid disagreements about its meaning.' If we consider just the measurement of space we could arrive at a range of different construals of the concept depending on who we ask. If we ask the surveyor they may express space in terms of net external area or net internal area. In contrast, the facilities manager may be more concerned with workplace area. Those occupying the space may be more preoccupied with the 'perceived' space in terms of feelings of overcrowding or openness. For the business manager, the emphasis is not on the physical measurement but on how the extent and location of the space impacts on power relations and process. The existence of these multiple meanings makes the challenge of building information management particularly difficult in that people assign different meanings to the same concept. Inevitably, these multiple definitions may or may not be found in the model. The net consequence of this is that effective sharing of information is hampered as different users have different requirements. This is particularly apparent in the two spheres of construction and facilities management, both of which possess widely different views of the physical asset. The temptation is to eliminate all ambiguities in such models. However, if we pursue the idea of 'human centred' approaches to FM then we would expect the information to be defined in such a way that it is useful to smaller units and/or specific stakeholders.

Conflict 2: Globalism versus particularism. In order to be of use, building information models need to be based on meanings that apply to an entire organisation or industry. This standardisation is reflected in the digital libraries of building components and systems. Whilst such libraries can represent many component types, the difficulty of encompassing the full range (reflecting differing national approaches and localised building practices) makes the idea of a unique 'engineered' component model problematic. Added to this is the idiosyncratic or localised nature of many facilities management operations (often arising from asset specificity, whereby building assets possess distinct characteristics). A tension thus arises between what might be described as *information globalism* and *information particularism*. Facilities management teams and individuals invariably seek to represent information in distinct ways that make sense to them in their particular setting.

Building information modelling presents many new questions and challenges. It provides an alternative to computer-aided facilities management (CAFM), which is largely data driven from a pragmatic approach, with information being provided from the bottom up. In contrast, BIM demands a whole life approach to information management involving cooperation between the design, construction and facilities professionals. If BIM is to be successful it needs to provide a mechanism for feeding information back to the design and construction phases (feedback) as well as transferring information at hand-over (feedforward).

7.5 Knowledge management

The amount of data generated by modern facilities is unprecedented. Intelligent buildings now generate copious data which is acted upon in real-time using control algorithms. The advent of building management standards and the arrival of 'open' networks has also enabled the integration of such feedback across a range of building functions. In addition, the almost unseen inclusion of embedded systems within building systems has given rise to a multiplicity of data sources and perhaps more importantly the capacity for a diversity of building functions to operate in synchronicity.

The above scenario describes an extreme of a spectrum of information flow that confronts a facilities manager. At the other end of the information spectrum is data that is not acted upon in real-time but that requires a significant amount of human judgement to translate data to useful knowledge. For such decision making, information technology represents only a 'tool' for managers. Perhaps one criticism of modern-day facilities management is that we have become overreliant on information technology whilst ignoring some of the key judgements that remain in the hands of the facilities manager. It is perhaps because of the diversity of data, information and knowledge that is presented to facilities managers that many of the key decisions can end up being neglected. In the subsequent part of this chapter we develop a 'systems' approach (viable systems model) that seeks to capture the whole picture, highlighting not only the possibilities of new technology but also the blind-spots. In so doing we isolate the key decision-making activities that remain for the facilities management team.

7.5.1 Translating data to knowledge

When does data become information? When does information become knowledge? In an environment of overwhelming data infusion answers to these question become pivotal for the facilities management team. We might, for example, consider *temperature readings* for a particular point in a work area to be 'data'. We might be alerted to the fact that the recorded temperature exceeds acceptable working temperatures, via our building management system. This constitutes 'information'. Taking it one step further, awareness that temperature may not significantly affect productivity or breach statutory regulations (when considered with the ambient air flow rate and duration) constitutes 'knowledge'. What is the key to effectively understanding the distinction between information and knowledge? Some would argue that information and knowledge differ in terms of context, usefulness or interpret ability. However, another common view is that 'knowledge is information possessed in the mind of individuals: it is personalised information (which may or may not be unique, new, useful or accurate), related to the facts, procedures, concepts, interpretations, ideas, observations and judgements' (Alavi and Leidner, 1998: p. 6).

Davenport (1994) argues that research since the 1960s has consistently shown that managers do not rely on computer-based information to make decisions. He contends that 'managers get two thirds of their information from face-to-face or telephone conversations; they acquire the remaining third from documents, most of which come from outside the organisation and aren't on the computer system' (Davenport, 1994: p. 121). During the time since Davenport made this observation technology has significantly changed. In particular, the advent of social networking has greatly increased the importance of the 'C' in ICT (information and communication technology), providing a panoply of dissemination routes, many of which now play an indispensable part in getting closer to the FM customer. These technologies enable rich communication between the information provider and information user. They reinforce established technologies like telephone communication that do not rely on the encoding of information (lean information) but instead rely on the personalised knowledge of the information provider and recipient. Such technologies can be synchronous (in other words, information exchange occurs simultaneously between sender and recipient) or alternatively asynchronous (rich information can be picked up by the recipient at a later point in time). Examples of rich media that enable synchronous exchange include videoconferencing (e.g. WebX or Skype) and IP telephony. Voice messaging, instant messaging and email are examples of asynchronous communication (they fall under the category of rich information as they facilitate bespoke, personalised exchange, although lacking the level of richness achievable using voice or video communication).

7.5.2 Knowledge management in FM

The field of knowledge management has emerged largely as a response to information technology's preoccupation with hardware and the need to redress the balance in favour of useful 'user-oriented' output. Facilities management has

Table 7.1   Examples of knowledge processes in facilities management.

Process	Facilities management example	IT/communications method
1. Generating new knowledge	• Customer feedback • Building monitoring	• Building management system (BMS) (Wang and Xie, 2002) • Helpdesk system (McBride, 2009)
2. Accessing valuable knowledge from outside sources	• Interorganisational benchmarking • Supplier monitoring	• e-procurement (Ancarani and Capaldo, 2005) • e-tendering (Ajam, Alshawi and Mezher, 2010)
3. Using accessible knowledge in decision making	• Procurement decisions • Building design decisions • Contractor selection • Risk assessment	• Automated scoring (e.g. analytical hierarchy process) (Gilleard and Yat-lung, 2004) • Digital prototyping (BIM) (Steele, Drogemuller and Toth, 2012; Shen, Hao and Xue, 2012) • Fuzzy logic in risk assessment (Tah and Carr, 2000)
4. Embedding knowledge in processes, products and/or services	• Configuring process flow • Optimising building management systems • Revising wayfinding and signage • Re-engineering service level agreements	Lean thinking and BIM (Sacks *et al.*, 2010) Lean and space planning (Durham and Ritchey, 2009) Continuous commissioning (Ahmed *et al.*, 2010)
5. Representing knowledge in documents, databases and software	• Developing a health and safety manual • Producing a workable business continuity plan	Electronic document management (Amor and Clift, 2006)
6. Facilitating knowledge growth through culture and incentives	• Developing a competency based training system	Knowledge management systems (Hansen, Nohria and Tierney, 2005)
7. Transferring existing knowledge into other parts of the organisation	• Roll-out of a corporate knowledge base system • Knowledge sharing hub between building managers	Collaborative virtual exchange and teleconferencing
8. Measuring the value of knowledge assets and/or impact of knowledge management	• Case studies • Competitions (internal and professional)	

been a particular beneficiary of this refocusing of IT efforts. One way of understanding knowledge management is in terms of a *process*. A four-stage process has been proposed by Alavi and Leidner (1998):

1. Knowledge creation (including knowledge maintenance and updating)
2. Knowledge storage and retrieval of knowledge
3. Knowledge sharing
4. Knowledge application

A more detailed breakdown of the knowledge management process has been proposed by Teece (1998) involving eight basic processes. Table 7.1 illustrates the type of information technology used to support each of these eight stages in the field of facilities management.

Whilst much attention has traditionally been given to explicit encoded data, FM's are increasingly aware of the importance of 'tacit' knowledge. This refers to the type of information and know-how that does not reside on a computer system. Instead, it is the critical information that exists in the minds of customers, contractors and others that have direct experiential learning. Rather than damping down such signals by means of database summary data, the IT strategy should seek to amplify their own capacity to deal with such variety. Crowd-sourcing technology sits at the heart of such an approach and relies on technologies such as social networking, instant messaging and online customer survey tools. Geiger, Rosemann and Field (2011) characterise crowd-sourcing as approaches that harness the potential of large and open crowds of people. A crowd usually emerges as an undefined network of people that responds to an open call for contributions to a particular task. In FM an application that has attracted particular attention is disaster recovery (Goodchild and Glennon, 2010). In relation to the capture of tacit knowledge using BIM tools, Barrett (2003: p. 253) argues for a 'broader, longer-term perspective beyond immediate project needs. Given the *tacit–tacit* emphasis of the industry, the mis-match with the *explicit–explicit* character of 4D CAD systems is stark. Instead of accuracy and detail, coarse robustness and connectedness are needed in systems that cover the important hard and soft dimensions.'

7.6   IT governance models

So far in this chapter, we have identified both the potential and the limitations of information technology in supporting the FM role. The arguments for using the latest technologies or contrariwise the arguments for sticking with a tried and trusted technology create an ongoing tension in organisations. The concept of a 'user-oriented' IT solution is particularly attractive in an environment where the sharing of information and knowledge cannot be taken for granted. Many examples exist of large-scale IT projects that have floundered as a result of information users' reluctance to engage with the technology and share information. Without careful engagement of information providers and users, resistance to such projects is almost inevitable. The barriers to information sharing extend beyond individuals. Jones and Dewberry (2012) point out the problems of what they call 'lonely and selfish' BIM, which continues to affect attitudes towards information sharing at the interorganisational level in the AEC industries.

7.6.1   Information flow under stress

Information technology strategies abound in organisations. What is typically missing is an overriding 'information strategy' in which technology plays only a part. Supporting the business needs of an organisation by means of responsive and cost-effective facilities demands IT solutions that are distinct from conventional business solutions. The integration of 'hardware' (building and building components) with the 'software' in the form of facility services makes specific demands on the IT system. Indeed, the information environment shares many characteristics with that of disaster response situations. These characteristics identified by Preece, Shaw and Hayashi (2012) include:

- High levels of uncertainty
- An environment of high stress with resulting high consequences
- Timelines that are significantly compressed
- Events for which there is an initial shortage of information followed by extreme information overload
- A multitude of actors (often with limited experience in working together)

High levels of uncertainty and unexpected events are far from exceptional in facilities operations. Indeed, IT systems in FM often only prove their worth in conditions of unanticipated upheaval. IT solutions in FM are often built around the assumption of a 'steady-state' condition. However, as well as the obvious scenarios of natural disasters (fires, floods, civil unrest) many other events such as major business discontinuities require rapid responses on the part of the facilities manager. Mergers, takeovers, relocations, the adoption of new business models, all of these present unexpected challenges for the facilities manager. Enterprise-wide CAFM data in the form of asset inventories, health and safety audits, space plans, condition surveys and other such databases inevitably play a key role in dealing with unexpected events. However, in such situations the facilities manager is dealing not only with *risks* that have emerged but also dealing with pure *uncertainty*. Unlike a risk that can be evaluated in terms of probability and magnitude, uncertainty presents a much more challenging scenario. Under conditions of uncertainty historic data may be of limited use. Knowing what has happened in the past may shed little light on what is going to happen in the future. It is the same with IT solutions in FM in coping with the increasingly uncertain environment (business, economic, statutory and social); dealing with change and uncertainty has become far more important than dealing with 'business as usual'.

7.6.2 *Defining IT governance*

Practical guidelines on the design of IT systems in facilities management as well as the selection of software systems have been produced by various professional bodies including the BIFM and IFMA. However, a holistic theory or model that can be used for IT strategy (or more precisely, information strategy) has yet to be put forward. A corporate IT governance model identifies the necessary elements required in formulating an FM IT strategy. The term 'governance of IT' is defined in ISO/IEC 38500 (2008) as:

… the system by which the current and future use of IT is directed and controlled. It involves evaluating and directing the plans for the use of IT to support the organisation and monitoring this use to achieve plans. It includes the strategy and policies for using IT within an organisation.

The term governance is derived from the Greek word *kyberman*, which means to be at the helm or to steer. This also links to the term 'cybernetics', which is concerned with the study of communications and control both in animals and machines. In modern-day parlance, the term cybernetics also encompasses the study of sociotechnical systems.

In the following section we explore a holistic model that satisfies the requirements of an IT governance model. We then explore how this can be applied to the facilities management context, enabling managers to develop a future-oriented IT approach that recognises the fundamental role of users and information providers.

7.7   Viable system model (VSM)

The idea of 'information architecture' is appealing in terms of predictability. The belief that we can structure in a rigid predetermined manner in the same way that we can design a house is somewhat beguiling. However, this mechanistic metaphor is perhaps less suited to the modern facilities management scenario of uncertainty and increased demands for user involvement. In the words of Davenport (1994: p. 122):

> … while information architecture can't capture the reality of human behaviour, the alternative is hard for traditional managers to grasp. That's because the human centred approach assumes information is complex, ever expanding and impossible to control completely. The natural world is a more apt metaphor for the information age than architecture.

Taking account of the decision-making environment encountered in FM, this section advocates the use of a governance model to steer and direct IT approaches in facilities management. Given the potential pitfalls arising from an incomplete use of IT tools, inappropriate use of IT tools, exclusion of users and inflexibility arising from a single rigid solution, a holistic governance model is seen as being indispensable.

The holistic model proposed is that of the 'viable system model' originally developed by Beer (1985), which forms the basis of the 'generic FM model' discussed in Chapter 1. The model itself has been used extensively and more recently has been used in defining governance models. In particular, Lewis and Millar (2009) have described a governance model based on VSM for use in corporate governance of IT. Preece, Shaw and Hayashi (2012) have used it to analyse information processing to aid disaster management. Stephens and Haslett (2009) developed a case study analysis for the purposes of strategic planning using action research based on the viable system model. The viable system model (VSM) of Beer (1985) provides a mechanism for coping with the spiralling complexity evident in the FM's external environment. To achieve an optimal response, the variety of the environment and the variety of the system needs to be balanced. In order to achieve such a balance an FM information system has to absorb increasing environmental variety either by increasing (amplifying) their own variety or by replacing (attenuating or 'damping down') the variety from the external sources. An example of an IT tool used in FM to attenuate signals occurs when the FM relies on totals and averages arising from aggregated data on contractor performance (e.g. using spreadsheet data). This invariably takes the form of 'encoded' knowledge, which can be objectively measured (statistically filtered). In contrast, crowdsourcing technologies, such as Facebook communities or LinkedIn communities, provide the opportunity to amplify diverse signals relating to changes in the external environment.

7.7.1   General systems theory

The 'viable system model' is based on the principles of 'general systems theory', which has been used extensively to model natural systems and social systems. Unlike models of mechanical systems it adopts a holistic perspective such that all information does not have to be common and some disorder and redundancy may be present and indeed may be desirable. Fundamental to this general theory are a number of key premises, which have been alluded to in Chapter 1. These include a recognition of (1) a system being composed of subsystems and components that are interconnected; (2) the system itself can only be explained in terms of its totality or as a holistic entity (the opposite of elementarism, a common assumption in many FM IT systems); (3) the system represents a transformative process involving various inputs, which are then transformed and exported as outputs; (4) closed systems tend towards entropy, which is a movement towards disorder, largely as a result of being out of touch with the changing external environment; in biological or social systems it is possible to arrest this decline as such open systems are capable of importing resources from their environment; (5) an open system is capable of attaining a state of dynamic equilibrium or a 'steady state' as a result of continuous flows of material, energy and information; (6) feedback plays a key role in maintaining a steady state including both positive and negative feedback arising from the outputs or the process of the system; (7) hierarchical relationships exist between systems and within systems; (8) biological and social systems involve multiple goals or objectives and facilities management addresses the needs of many individuals and subunits, each with different values and objectives; (9) unlike mechanistic physical systems in which we can identify direct cause and effect relationships concerning initial conditions and the final state, biological and social systems are very different. This leads to the concept of equifinality, which suggests that it is possible to achieve the same result through different means. From the facilities management point of view it suggests that it is possible to achieve the same objective (e.g. satisfied users) using a variety of different inputs and using different conversion processes.

7.7.2 Five key concepts in VSM

How do we design a viable IT solution in facilities management? The key to 'viability' is the capacity of a system to 'maintain its identity independent of other such organisms within a shared environment'. A viable system is thus able to maintain its identity independently within an environment that is complex and involves multiple users, suppliers, competitors, etc. It is thus able to survive despite significant changes from within and in the external environment. It is suggested that such a viable (VSM) structure also leads to greater effectiveness with non-viable organisations encountering crises that viable organisations would not be subject to. It is the ability to model information flow and communication links that such a tool provides. In the facilities management context it enables the facilities manager to highlight existing or missing communication patterns and information flows that arise in different communication channels. The organic nature of VSM enables us to overcome the inflexibility of hierarchical structures that often beset facilities management structures and present a barrier to information sharing.

This leads to the five key concepts of VSM which include the following.

Viability

A viable system is one capable of surviving in a world that is increasingly competitive and complex. This is achieved through a capacity to learn, adapt and grow, just like a living organism. Moving beyond simply the idea of survival, viability implies the ability to thrive and grow through value creation.

Variety

Fundamental to VSM is the 'management of complexity' within the organisation. Linked to this is 'Ashby's law of requisite variety', which states that only variety can destroy variety. In previous chapters we have seen that the pursuit of 'commoditisation' in facilities management is a common driver in the pursuit of reduced costs and consistent quality. However, set against this is the ever-changing external environment, which throws up variety in many shapes and forms. The implication of Ashby's law is that facilities management organisations must possess sufficient variety to enable them to match the environmental states that may challenge their 'steady-state' condition.

Recursion

Embedded within a facilities management department would be a number of sub-units, including, for example, maintenance management, security, hospitality and health and safety. Each of these may reside in different building portfolios located in different countries. A viable system such as a facilities management department would itself be contained in a viable system such as the organisation and indeed within an industry, which itself would be viable. This gives rise to what is described as recursion in the VSM model whereby systems are contained within systems.

Autonomy

We are used to the problem of different organisational units within the organisation pursuing different goals. In order to prevent such 'component parts' going off in their own direction, mechanisms are put in place by the meta-system to prevent just such an occurrence. The VSM model thus distributes authority between organisational levels with a view to maintaining organisational cohesion. Whilst a certain degree of autonomy is required to create, regulate and implement particular policies, these need to be aligned with the organisation as a whole. In facilities management we see examples of this in space planning whereby a rigid space standard is replaced by a space template that can be adapted or modified at a local level to suit individual or department needs. However, constraints or limits define a more 'elastic' boundary around such a decision-making process.

Transduction

One of the key skills of a facilities manager is the ability to speak different languages according to the audience. When looking at communication links in a VSM model, a key requirement is that whenever communication occurs across a boundary, the information must be 'translated' into the 'language' of the receiving subsystem. If we look particularly at reporting systems used in facilities management IT, reports directed at the Board of Directors need to be translated using a language fundamentally different from that of technical staff or building users.

7.7.3 The VSM model explained

The anatomy of the VSM model is outlined in Figure 7.2 and is proposed as a 'governance' model for use in facilities management. It resembles the generic FM systems model introduced in Chapter 1 (see Figure 1.1), but also encompasses several concepts that specifically apply to information flow in IT. The model itself comprises five discrete systems or functions that are connected via a series of information flows. Shown on the left-hand side of Figure 7.2 is the environment encompassing both the 'future' state and the 'current' state, as previously outlined in Chapter 1.

Considering the system from the bottom up, we begin with System 1, which represents the various operational units ($S1_a \ldots S1_n$) within the FM organisation as well as other organisational units with which they interact. Each of these interact with their own 'local' environments, which may be specific to their geographical location or area of specialisation (e.g. maintenance management) connecting with System 1. The lines shown between System 1 and Systems 2 to 5 represent information flows. Also included is a subsystem of System 3 represented by System 3*, which embraces the Audit/Monitor function.

System 1: Operations. Being concerned with the day-to-day operation of facilities, each of these organisational subunits are themselves self-contained viable systems. They possess the ability to self-organise and self-regulate, and indeed

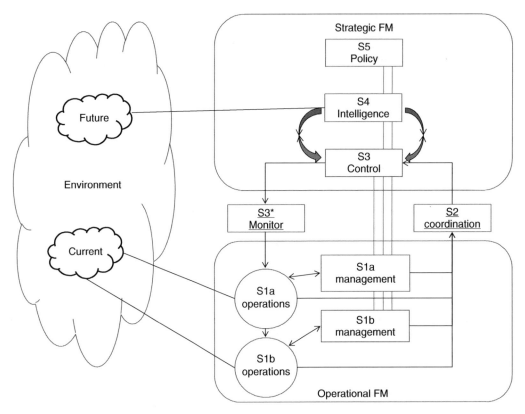

Figure 7.2   Viable systems model (VSM) based on Beer (1985). © 1985 John Wiley & Sons. Reproduced with permission of John Wiley & Sons.

the governance model provides freedom for the individual units to act autonomously. This autonomy thus reduces the amount of variety that the meta-system (i.e. Systems 3 to 5) has to deal with.

System 2: Coordination. Is responsible for ensuring that the primary activities of the subunits in System 1 (operational level) do not conflict with one another such that each of the different teams avoid duplication.

System 3: Control. Also described as 'cohesion', this system oversees the control of System 1 elements as part of an overall control function in a holistic way. An 'integrated' way implies one that involves minimal intervention by the meta-system consistent with the cohesiveness of the entire viable system.

System 3*: Monitoring. Without this additional monitoring component, System 3 presents a highly filtered communication channel for controlling System 1. To address this limitation of highly filtered information an additional system (System 3*) is required to accommodate the high levels of variety that inevitably arise in System 1 at the operational level.

System 4: Intelligence. To be viable the system must be able to not only control its own environment by addressing the 'here and now' but must also have the capacity to adapt to changes in the external environment that may undermine its own viability. This environmental scanning enables the system to contemplate the future and addresses 'outside and then'. This is shown in Figure 7.2, which shows the close connection of information flow in the form of scanning between the external environment and the intelligence system.

System 5: Policy. This part of the overall system represents the 'guiding mind' behind the viable system. When looking at the design of an IT system for facilities management, such a guiding framework is essential. As part of this, processes and activities that run alongside the necessary information are defined. Examples of overriding philosophies include:

- Environmental assessment methods (e.g. BREEAM, LEED, Green Star).
- Lean thinking/Six Sigma (concerned with maximising customer value whilst minimising waste). For further information see http://www.lean.org/.
- PAS 55 (the British Standards Institution's (BSI) Publicly Available Specification for the optimized management of physical assets based on a 28-point requirements specification). For further information see http://pas55.net/.
- Soft Landings (concerned with keeping contractors and designers involved with a facility post-completion, a methodology that seeks to ensure that the occupiers understand how to control and best use their buildings). For further information see http://www.bsria.co.uk/services/design/soft-landings/.

7.7.4 Mapping of FM technologies

Being able to model 'information flows and communication links' (Flood and Jackson, 1991: p. 92) is a fundamental starting point for any FM blueprint in choosing appropriate technologies. The word 'information' precedes the word 'technology', both in terms of order and importance. Technologies do not necessarily have to be 'leading edge' in order to justify their inclusion in such a blueprint. This is evidenced by the profound impact of barcoding in the retail industry, which continues to rely on a simple linear barcode solution based upon a universal retailing standards for item identification (the Universal Product Code) formulated over four decades ago. Despite the appearance of many competing and seemingly more advanced solutions to the problem of item identification (including two-dimensional bar-coding and radio-frequency identification), the traditional barcode continues to outlive its expected life. The robustness and the longevity of the technology owes much to the commitment of global retailers to a single standard. Much could be said about technologies in facilities management. It often does not matter what technology is chosen provided that there is a common agreement between parties involved in the FM enterprise regarding its choice and implementation.

In order to 'highlight existing or missing communication patterns and information flows in different communication channels' (Nyström, 2006: p. 523) the following attempts to map FM technologies to the viable system model. Table 7.2 illustrates a few of the systems commonly used in FM, breaking down elements of the systems into the five constituent parts of the viable system model.

Table 7.2 Mapping of FM technologies in the VSM model.

Viable system	FM application		
	Energy and sustainability	Space and workplace management	Maintenance management
System 5: Policy	Building information modelling BREEAM, LEED, Green Star Soft Landings	Building information modelling User-oriented FM Organisation design (OD) Lean thinking	Building information modelling PAS 55 Lean thinking ISO 9000; ISO 14000
System 4: Intelligence		Demographic trends analysis	Flood level monitoring
	Benchmarking Modelling (BIM) Geographic Information Systems	Business forecasting RFID asset tracking Crowdsourcing	Fleet management
System 3: Control		Timetabling systems	Resource scheduling systems
	Building energy management system Intelligent Buildings	Access control	
System 3*: Monitoring	Condition monitoring Fire detection	Occupancy monitoring	Condition monitoring
System 2: Coordination	Network enabled embedded systems (Internet of things)	Location tracking (RFID, near-field and bar-code technologies)	Network enabled embedded systems (Internet of things)
System 1: Operations	Remote sensors Embedded systems Optimisation and load balancing	Kiosks Employee self-service	Automated identification (RFID, barcoding) Embedded systems Motes GPS tracking

7.8 Conclusions

This chapter has explored the possibilities as well as the limitations of information technology in facilities management. It advocates a proactive approach by the facilities management team in formulating an IT strategy. It proposes the use of the viable system model as a way of dealing with the necessary variety that is required in a 'user-oriented' approach. Major developments in software architecture (in particular building information modelling) present a growing opportunity for harnessing information that in the past has been lost in the construction phase. However, facilities management is concerned with much more than the physical building. In order fully to exploit technologies such as BIM that are derived from a singular view of the construction process, a more complete portrayal of facilities management is necessary. This entails capturing the multi-faceted views of facilities management as seen by different stakeholders. In order to be fully successful, a facilities management IT strategy must identify elusive and complex communication patterns in a variety of communication channels. Some of these channels will take the form of 'encoded' knowledge that can be expressed within a BIM or CAFM model. However, much of the requisite information will remain 'unencoded' and the challenge for the facilities manager is to remain sensitised to such information.

The next chapter follows on to consider decision making in facilities management. One clear message that emerges from this chapter on IT is that, despite the abundance of data and information, knowledge remains a scarce resource. Invariably it falls to the facilities manager to make sense of varied and complex information in order to arrive at a decision. No amount of information technology can supplant or obviate this key role.

References

Ahmed, A. et al. (2010) Multi-dimensional building performance data management for continuous commissioning. *Advanced Engineering Informatics*, **24**(4), 466–475.

Ajam, M., Alshawi, M. and Mezher, T. (2010) Augmented process model for e-tendering: towards integrating object models with document management systems. *Automation in Construction*, **19**(6), 762–778.

Alavi, M. and Leidner, D.E. (1998) Knowledge Management and Knowledge Management Systems: Conceptual Foundations and an Agenda for Research. INSEAD.

Amor, R. and Clift, M. (2006) Document Management in Concurrent Life Cycle Design and Construction. Also available from Taylor & Francis, p. 183.

Ancarani, A. and Capaldo, G. (2005) Supporting decision-making process in facilities management services procurement: a methodological approach. *Journal of Purchasing and Supply Management*, **11**(5), 232–241.

Barrett, P. (2003) Construction Management Pull for 4D CAD. Construction Congress VI, American Society of Civil Engineers, pp. 977–983.

Barrett, P. (2005) Revaluing Construction: A Global CIB Agenda. International Council for Research and Innovation in Building and Construction (CIB), Rotterdam, the Netherlands.

Beer, S. (1985) Diagnosing the System for Organizations. John Wiley & Sons, Ltd, Chichester.

Davenport, T.H. (1994) Saving IT's soul: human-centered information management. *Harvard Business Review*, **72**(2), 119–131.

Durham, J. and Ritchey, T. (2009) Leaning forward. *Health Facilities Management*, 23–27.

Flood, R.L. and Jackson, M.C. (1991) Total systems intervention: a practical face to critical systems thinking. *Systems Practice*, **4**(3), 197–213.

Gallaher, M.P. and Chapman, R.E. (2004) Cost Analysis of Inadequate Interoperability in the US Capital Facilities Industry. US Department of Commerce, Technology Administration, National Institute of Standards and Technology.

Geiger, D., Rosemann, M. and Fielt, E. (2011) Crowdsourcing Information Systems: A Systems Theory Perspective. *Proceedings of the 22nd Australasian Conference on Information Systems (ACIS 2011)*.

Gilleard, J.D. and Yat-lung, P.W. (2004) Benchmarking facility management: applying analytic hierarchy process. *Facilities*, **22**(1/2), 19–25.

Goodchild, M.F. and Glennon, J.A. (2010) Crowdsourcing geographic information for disaster response: a research frontier. *International Journal of Digital Earth*, **3**(3), 231–241.

Hansen, M.T., Nohria, N. and Tierney, T. (2005) What's your strategy for managing knowledge? *Knowledge Management: Critical Perspectives on Business and Management*, **77**(2), 322.

International Standards Organisation (2008) IT Governance Standard ISO38500.

Jones, D. and Dewberry, E. (2012) Building Information Modelling Design Ecologies – A new model? Available at: http://oro.open.ac.uk/34329/ (accessed October 5, 2012).

Lewis, E. and Millar, G. (2009) The Viable Governance Model – A Theoretical Model for the Governance of IT. In IEEE 42nd Hawaii International Conference on System Sciences, HICSS'09, January 2009, pp. 1–10.

McBride, N. (2009) Exploring service issues within the IT organisation: four mini-case studies. *International Journal of Information Management*, **29**(3), 237–243.

NBS (2012) National BIM Report 2012. Available from: http://www.bimtaskgroup.org/.

Nyström, C.A. (2006) Design rules for intranets according to the viable system model. *Systemic Practice and Action Research*, **19**(6), 523–535.

Preece, G., Shaw, D. and Hayashi, H. (2012). Using the viable system model (VSM) to structure information processing complexity in disaster response. *European Journal of Operational Research*.

Sacks, R., Koskela, L., Dave, B. and Owen, R. (2010) Interaction of lean and building information modeling in construction. *Journal of Construction Engineering and Management*, **136**(9), 968–980.

Shen, W., Hao, Q. and Xue, Y. (2012) A loosely coupled system integration approach for decision support in facility management and maintenance. *Automation in Construction*, **25**(0), 41–48.

Steel, J., Drogemuller, R. and Toth, B. (2012) Model interoperability in building information modelling. *Software and Systems Modeling*, **11**(1), 99–109.

Stephens, J. and Haslett, T. (2009) The Application of Stafford Beer's Viable Systems Model to Strategic Planning. In *Proceedings of the 53rd Annual Meeting of the ISSS*.

Tah, J.H.M. and Carr, V. (2000) A proposal for construction project risk assessment using fuzzy logic. *Construction Management and Economics*, **18**(4), 491–500.

Teece, D.J. (1998) Capturing value from knowledge assets. *California Management Review*, **40**(3), 55–79.

Wang, S. and Xie, J. (2002) Integrating building management system and facilities management on the Internet. *Automation in Construction*, **11**(6), 707–715.

8 Decision Making

8.1 Introduction

8.1.1 The importance of decision making

Decision making is an integral part of the facilities manager's role. Facilities managers have to process information continuously and make decisions concerning all apects of the work environment. For example, the information produced from a post-occupancy evaluation should be analysed and fed back into the decision-making process for future space-planning policies and so on. Cumulatively, such decisions plan, organise and control an organisation's facilities so that they support the organisation's primary business needs.

Managers generally concentrate on decision *output*. For example, the decision to install a building management system is often assessed on such variables as its impact on fuel costs. However, such a preoccupation with assessing the decision output tends to underplay the role of how decisions are made. It can be readily appreciated that the *effectiveness* of decisions is determined predominantly by the quality of the decision-making *process* used to generate it. For instance, decisions geared towards controlling occupancy costs are only as good as the process used to collect and analyse the occupancy cost information. Improvement in the decision-making process is therefore a very good opportunity for facilities managers to consistently improve the decisions they make.

8.1.2 The myth and reality of decision making

In general, much is assumed but little is known about this important managerial activity. The opinions of facilities managers regarding their own decision-making abilities are conditioned strongly by what is considered a satisfactory decision to be. In consequence, when it is suggested to facilities managers that they might be able to improve decision-making techniques, they often respond with a highly defensive reaction. For example, who among the readers of this chapter would admit that they are not a good decision maker?

Facilities Management: The Dynamics of Excellence, Third Edition. Peter Barrett and Edward Finch.
© 2014 John Wiley & Sons, Ltd. Published 2014 by John Wiley & Sons, Ltd.

However, it has been observed that decision making is a complex, irrational process. Decision makers generally:

1. *Apply few special decision-making procedures when arriving at their choice*: for example, facilities managers often do not use a systematic procedure to assess the impact of an office-layout change upon related building elements, such as raised floor systems, floor service outlets and air conditioning systems.
2. *Lack of information about the merits and consequences of alternatives*: for example, in matters of facilities planning and furniture allocation, facilities managers often fail to consider alternatives. Instead, facilities managers apply predetermined organisational standards without question. Such a stance runs the risk of alienating individual user needs, potentially generating all the problems associated with resistance to change. (See Chapter 5 on Managing People Through Change and Chapter 3 on Engaging with Stakeholder Needs for further discussion on the importance of participation.)

8.1.3 *The need to rationalise the decision-making process*

In spite of the apparently chaotic nature of the decision-making process, it is essential to the overall success of the facilities management function that its managers develop more rational decision-making procedures and generally strive to improve their decision-making capabilities. Experience has demonstrated that being more rational improves managerial decision making of all types and considerable benefits can be reaped, including:

- providing more structure to poorly structured problems;
- extending the manager's information-processing ability;
- providing cues to the managers of the critical factors in the problem, their importance and the relationships between them;
- breaking out from 'blinkered' frames of mind to view problems from new perspectives.

For example, facilities managers can provide a more *structured* base from which to make better energy-management decisions if they have the benefit of systematic procedures for the collection, recording and evaluation of energy-use information. By providing timely and useful information, facilities managers can *process* more information, as they are not hampered by first having to sort out the relevant from the irrelevant information. As the information is more focused to the facilities managers' needs, they are in a better position to identify the *critical factors* and observe trends more readily. For instance, fuel-consumption trends can indicate that heating and ventilation plant require maintenance. Finally, the facilities manager will be in a better position to view occupancy costs from a proactive, rather than a reactive, *perspective*. Such a perspective for energy management can allow greater accuracy in scheduling preventive maintenance and thus reduce costly equipment failures.

However, facilities managers have been known to dismiss any rationalisation of the decision-making process. Instead, managers frequently maintain that experience alone is sufficient to achieve good decisions. It is suggested that such reasoning is a dangerous path to follow, as:

> … past experience in decision making is no guarantee that our experiences have taught us the best possible methods of … decision making and problem solving. Learning from experience is usually random.
>
> Furthermore, although we all learn experiences, there is no guarantee that we learn *from* experiences. In fact, it is possible to learn downright errors and second-rate methods from experience, as in playing golf without taking lessons from a professional. As with the golfer, so with the manager: it is only training in systematic method which enables us to correctly analyze situations so that we can truly learn from experience in those situations (Elbing, 1970: p. 14).

This is not to dismiss experience as worthless. Experience has its own essential role to play within the rationalised decision-making structure. Moreover, facilities managers should view decision making as a process that can be improved by working on integrating rational decision making with their own *intuitive* and *commonsense* approach to decision making.

Indeed, this mix of a rational step-by-step approach and intuition can be essential when the problem being addressed is rather amorphous and has multiple dimensions within any particular solution. This is compounded by situations where shared objectives only become clear as a result of the process, rather than providing a touchstone for it. The majority of this chapter emphasises a rational decision-making approach towards a clear solution. However, at the end a case study is provided of a structured, but more open-ended, approach, with the emphasis on creating a clearer joint understanding of the problem situation.

8.1.4 *The decision-making process*

In developing structured approaches to the decision-making process, the facilities manager should appreciate the relationship between the *decision-making* process and the *problem-solving* process. In this chapter, the decision-making process will be considered as *part* of the larger process of problem solving. The decision-making process focuses around the managerial tasks of sensing problems and choosing between possible solutions. Problem solving is a broader process that includes the implementation of the solution, along with the solution's follow-up and control. Figure 8.1 shows the stages of the problem-solving process and separates those stages that form the subprocess of decision making and are largely the concern of this chapter.

The decision-making process begins with the exploration of the nature of the problem, continues through the generation and evaluation of possible solutions, culminating in the choice of an option. For the purposes of simplification, the chapter will consider each element as a self-contained phase. It will be readily

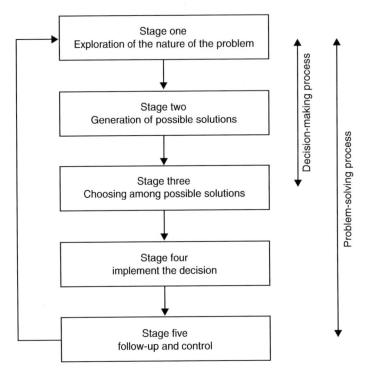

Figure 8.1   The basic model of the problem-solving process.

appreciated by the reader, however, that in reality there is not a simply sequential relationship between them. The predominant objective of delineating distinct phases is to help ensure that the facilities manager has considered the most relevant factors and all necessary actions have been taken. In addition, it is hoped that it will enable readers to enter the decision-making process easily at whatever stage they wants to for a given problem situation. However, facilities managers should appreciate that such an approach only states what they *should* do; it does not specify *why* the problem should be solved or its priority within the organisation. This is where experience and knowledge of the unique situation comes in.

8.1.5   Chapter structure

Each of the separate decision-making process *stages* will be discussed in turn in the remainder of this chapter, with the use of relevant theory, managerial tools and case study material. Each stage has been split into *steps*, which in turn have been examined under a standard format of *sections*. These sections are:

- *Step objectives*. The purpose of the step.
- *Step rationale and context*. How the step fits into the overall decision-making process.

- *Tasks*. The individual tasks to be completed within the step.
- *Tools*. Managerial techniques that can be used to accomplish the tasks. These tools are intended to give ideas only. It is accepted that they may be modified or rejected depending on the organisational situation and the nature of the decision.
- *Input*. The information required to carry out the step.
- *Output*. The tangible product of the step.

8.2   Stage 1: Exploration of the nature of the problem

8.2.1   Introduction

The objective of this stage is to provide the remainder of the decision-making process with a sound foundation, reducing the risk of generating an inappropriate solution and/or excessive use of organisational resources. In essence, this exploration stage gives overall direction and builds in the potential for added value where the benefits of the *outcome* of the decision-making process exceed the required *input* of organisational resources. Figure 8.2 shows the steps in the exploration of the problem stage.

8.2.2   Step 1: Sense problem

Step objective

- To detect problems effectively and efficiently.

Step rationale and context

Problem sensing is where managers detect a *problem gap* between a present situation and a desired situation (see Figure 8.3). Managers usually detect problems when:

- there is a deviation from past experience;
- there is a deviation from a set plan;
- other people present problems to them;
- competitors outperform their organisations.

For example, the decision to relocate may be triggered off in response to the need to provide expansion space, to check rising occupancy costs or because of changing markets in which the organisation operates.

The problems detected from these sources can be viewed as falling along a continuum. At one end there are *opportunity* problems, whose solution is initiated on a voluntary basis to improve an already secure situation. At the other end are *crisis* problems, where a situation arises that requires immediate attention. An obvious example of this is the relationship between planned and day-to-day maintenance.

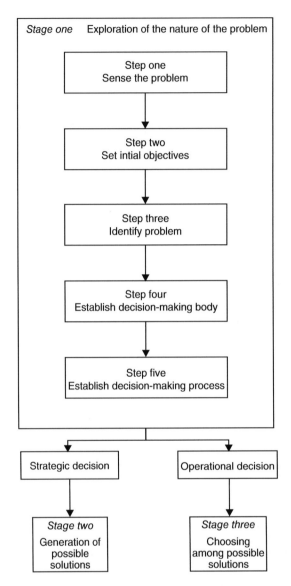

Figure 8.2   Exploration of the nature of the problem stage.

Planned maintenance can be viewed as the opportunity end of the continuum, while the day-to-day maintenance can be seen as the crisis end of the continuum.

With the rapid change being experienced by organisations, there is an increasing tendency for problems to be detected only towards the crisis end of the continuum. The aim for facilities managers is to develop problem-sensing mechanisms that will enable them to detect problems early, so that the problem is nearer the opportunity end of the continuum. This will give the facilities managers more time to come up with high-quality solutions. To come back to the

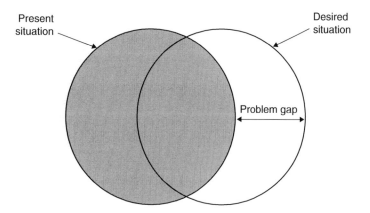

Figure 8.3   The problem gap.

example of planned versus day-to-day maintenance, it is generally the aim of all facilities managers to place increasing emphasis on the planned maintenance function in pursuit of medium- to long-term cost benefits.

Task

- Monitor the organisational environment for problem gaps.

Tools

Environmental scanning techniques
The facilities manager will generally have scanning techniques that survey the external and internal environment for potential problems. Optimal scanning techniques should aim to:

- interpret information quickly and in a meaningful way;
- maximise the information yield from the information that is collected, and thus minimise the costs of the scanning technique.

For example, an increasingly common way for facilities managers to effectively and efficiently sense problems is to employ benchmarking techniques. This process of comparing an organisation's facilities performance with other similar organisations can be a positive, structured method of identifying and assessing an organisation's strengths and weaknesses.

Input

The input required for this stage is the output from a management information system (MIS). It is suggested that any MIS should embody a critical success factor approach to enable the establishment of performance indicators. This is where an MIS is developed around strategic control points that are considered essential for the successful running of the facilities management function. This allows facilities managers to reduce the potentially overwhelming amount of information available to them to a subset of information that monitors the most crucial areas only. By streamlining information acquisition in this way, the facilities manager can be more perceptive to the more critical problem gaps as they occur.

Output

- A detected problem gap.

8.2.3   Step 2: Set initial objectives

Step objective

- To indicate what would constitute an effective and efficient solution.

Step rationale and context

The setting of objectives provides the end point towards which the decision-making process can be channelled. When setting objectives, the facilities manager should ensure that there is:

- a criterion against which the desirability of subsequent courses of action can be measured;
- a standard for evaluating performance once the decision has been implemented;
- the basis for self-evaluation by the decision maker on personal performance, acting as a benchmark for improvement for future decisions;
- flexibility within the decision-making process, to the extent that it is able to respond to ongoing changes.

In the absence of objectives, the decision-making process will tend to produce solutions that are without focus and, more likely than not, at cross-purposes with the general direction of the organisation.

The importance of coherence between higher and lower objectives was illustrated when the London Underground facilities management decided to introduce a computer-aided drawing system. The organisation, in response to cuts in funding by the Government, had set the *primary objective* of recovering the cost of space more effectively. In line with this policy, the facilities management decided to computerise the property-management system, with the medium- to long-term aim of introducing a charging policy for space occupation. However, the *initial objective* set was to ensure that the CAD system satisfied the primary objective of recovering space cost, through improving space efficiency via greater accuracy in planning.

Task

- Formulate objectives that are specific, verifiable and attainable.

Table 8.1   SMART checklist.

Decision objective checklist	Date	
Problem description		
Proposed objective		
	Problem characteristics	**Comments**
Specific. Is the proposed objective sufficiently clear to avoid ambiguity and uncertainty?		
Measurable. Does the proposed objective enable its performance to be evaluated?		
Attainable. Is the proposed objective realistically attainable?		
Relevant. Is the proposed objective consistent and linked to other organisational objectives and processes?		
Trackable. Does the proposed objective enable progress towards its accomplishment to be monitored?		

Tool

- Decision objective checklist.

A fruitful method for generating good-quality objectives is to consider proposed objectives against a SMART checklist (see Table 8.1). The facilities managers, by working through the checklist, can carry out a critical analysis of the usefulness of any objectives proposed. A worked example is shown later in Table 8.10.

Input

Relevant information is required about the problem and its environment in order to formulate initial goals. For example, a facilities manager, when deciding on a furniture policy, may work towards the general objective of optimising the organisation's accommodation assets to meet the organisation's business needs, over time, in the most cost-effective fashion. However, in order to concentrate this broad goal into SMART objectives, information may be required from:

- *the external environment*, say from a benchmarking exercise, to assess what other similar organisations are doing;
- *the external/internal environment boundary*, such as the likely organisational regroupings, with their associated churn, in the short to medium term;
- *the internal environment*, say existing organisational performance standards on shelving, screens, socket outlets, pedestals and work surfaces.

Output

- Defined solution objectives.

8.2.4   Step 3: Identify problem characteristics

Step objective

- To correctly identify whether a sensed problem is strategic or operational.

Step rationale and context

Once the facilities manager has sensed that there is a problem, the next task is to categorise the problem. The way in which problems are categorised affects subsequent decision-making performance. Perhaps the most important variation between problems is whether the problem requires a *strategic* or *operational* decision for its solution.

Strategic decisions are concerned with matters that relate the facilities management function to the external business environment. They tend to be long term in their effects and direct the facilities management function towards the maintenance of primary objectives. Strategic decisions tend to be ill-structured in nature, with the uncertainty and ambiguity created by the external business environment making them complex and open-ended. In contrast, *operational decisions* interpret a facilities management function's strategic objectives into manageable terms for short-term decision making. Operational decisions tend to be well-structured, repetitive and routine, with specific and concise organisational procedures formulated for handling them.

The significance of differentiating between these two basic categories of decisions became apparent, for example, when an organisation made a review of the provision of drinks within its headquarters. Up to this point a tea-trolley service had always been used, but it was assessed that the service cost approximately £25,000 per annum. This traditionally operational matter was elevated to that of a strategic problem, with concern being expressed over such issues as:

- Was the present arrangement good value for money?
- How did the service's cost and quality compare with its competitors?
- Could organisation-wide lessons be learnt from this problem?

The facilities manager, in this particular case, was faced with uncertainty over the effectiveness and efficiency of providing a vending service, in addition to being concerned over the social impact on the staff of stopping the tea-trolley service.

Another example is an organisation that has recently decided to change an earlier decision to outsource its catering function and keep it in-house. This turnaround was initiated by the facilities management which, upon reflection, decided that the catering function was of strategic importance, due to the nature of the primary business, and could be a source of competitive advantage if it was kept in-house.

Table 8.2   Decision-type diagnostic checklist.

Decision-type diagnostic checklist			
Problem description			
Problem characteristics	**Operational**	⇔	**Strategic**
Rarity. How frequently do similar problems occur?	Not rare Notes:		Very rare
Radicality of consequences. How far is the solution of the problem likely to change things within the organisation?	Not radical Notes:		Very radical
Seriousness of consequences. How serious would it be for the organisation if the chosen solution of the problem went wrong?	Not serious Notes:		Very serious
Diffusion of consequences. How widespread are the effects of the decision likely to be?	Not widespread Notes:		Very widespread
Endurance of consequences. How long are the effects of any decision likely to remain?	Not long Notes:		Very long
Precursiveness. How far is the solution of the problem likely to set parameters of subsequent decisions?	Not precursive Notes:		Very precursive
Number of interests involved. How many parties, both internal and external to the organisation, are likely to be involved in the solution of the problem?	Few parties Notes:		Many parties
Summary			

Task

- Define the problem as being strategic or operational in nature.

Tool

- Decision-type diagnostic checklist.

A useful way for the facilities manager to differentiate operational from strategic decisions, is to consider the perceived problem against the checklist shown in Table 8.2. A worked example is shown later in Table 8.11.

Input

Relevant information is required about the sensed problem to complete the decision-type diagnostic test. In the case of the outsourcing problem, information was needed to determine the relationship between the catering function and the core strategy of the organisation. In this particular case, the facilities manager had to collect information on the present and future nature of the core business of his organisation, along with information concerning the present and projected role of the catering function.

Output

- The problem is defined as being either operational or strategic. If the problem is strategic, go on to the next step. If it is operational, go on to Stage 3: Step 1A.

8.2.5 *Step 4: Establish a decision-making group*

Step objective

- To establish the optimum decision-making group with respect to the nature of the problem and the organisational situation.

Step rationale and context

Once the facilities manager has sensed and defined a strategic problem, the next stage is to determine the *optimum decision-making group* with respect to the type of problem that is being confronted. It is suggested that a productive way of determining the most appropriate decision-making group is to decide on the level of *participation* to be given to others in the decision-making process. Facilities managers can be *autocratic* in nature, making decisions within their area of authority, issuing orders to people and monitoring their performance to ensure compliance with their instructions. For example, when planning an office environment, facilities managers can assume the role of the expert, by virtue of their educational training and understanding of people, and design an environment that will satisfy the people and their needs. This approach is based on the notion that it is undesirable for the eventual users to participate in the planning of the environment, since they get in the way and do not have the necessary experience. Furthermore, participatory decision making makes the project much more expensive and time consuming.

In contrast, the facilities manager can provide for opportunities for those impacted by a decision to *participate* in the decision-making process, forming groups in order to share problems with them and encouraging them to arrive at mutually agreed solutions to problems. In the case of planning an office environment, the facilities manager can stimulate participation in the belief that people need to participate in planning their own environment in order to be satisfied. Through participation in the decision-making process, users have a feeling of control over their environment and it is the only way users' values can really be taken into account.

The reader will readily appreciate that these two contrasting decision-making styles form a *participation continuum*. It is useful to break down this continuum into five alternative managerial decision-making styles:

- *Autocratic 1 (A1)*. The managers make the decision themselves using information available to them at the time.
- *Autocratic 2 (A2)*. The managers obtain necessary information from others, then take the decision by themselves. The manager may or may not tell the others what the problem is when getting the information from them. The role of the others is purely that of providers of information, rather than that of generators or evaluators of alternative solutions.

- *Consultative 1 (C1)*. The managers share the problem with relevant people individually, getting their ideas without bringing them together as a group. The managers then make the decision, which may or may not reflect the others' inputs.
- *Consultative 2 (C2)*. The managers share the problem with others as a group, collectively obtaining their ideas and suggestions. The manager then makes the decision, which may or may not reflect the others' influence.
- *Group 1 (G1)*. The managers share the problem with other people as a group. The group generates and evaluates alternatives and reaches a mutually agreed solution. The manager does not try to influence the group and is committed to accepting and adopting any solution that has the support of the whole group.

The task for the facilities manager is to identify which of those styles is appropriate for a given situation.

Task

- Conduct a situational analysis of the problem.

Tool

- Situational leadership analysis.

The facilities manager can select the most suitable level of participation within the decision body for a given problem by using a checklist. The checklist comprises seven 'problem attribute' questions, each of which requires a simple Yes/No response.

The questions are as follows:

- *Question A*. Is there a quality requirement such that one solution is likely to be preferable to another?
- *Question B*. Do you have sufficient information to make a high-quality decision?
- *Question C*. Is the problem operational in nature?
- *Question D*. Is the acceptance of the decision by others critical to effective implementation?
- *Question E*. If you were to make the decision by yourself, is it reasonably certain that it would be accepted by others?
- *Question F*. Do others share the organisational goals to be obtained in solving this problem?
- *Question G*. Is conflict among others likely for the proposed solution?

By answering these questions, the facilities manager can diagnose a situation fairly quickly and accurately, and generate a prescription concerning the most effective decision body.

The simplest way to diagnose the appropriate decision body is through the use of a decision tree, as shown in Figure 8.4. The questions A to G (above) are

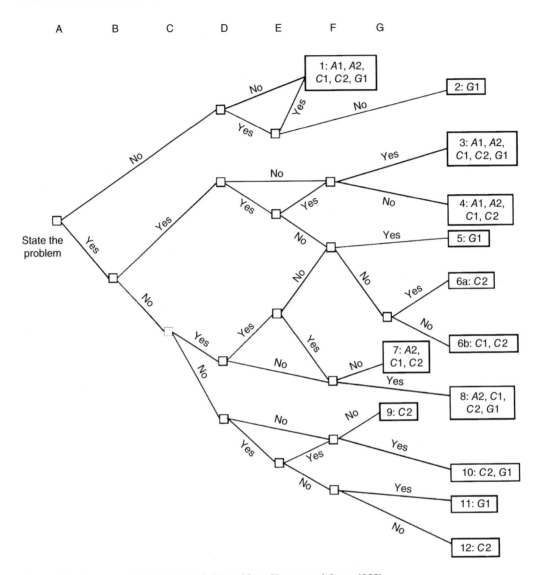

A   B   C   D   E   F   G

Figure 8.4 Decision-body flowchart (adapted from Vroom and Jago, 1988).

arranged at the top of the figure. To use the tree, the facilities manager selects the problem that has been sensed, enters the decision tree on the extreme left-hand side at State the Problem and asks the first question: Does the problem possess a quality requirement? The answer, yes or no, denotes a path that leads to another question identified by the letter immediately above the box. The process continues until the facilities manager encounters a terminal node (i.e. an endpoint on the tree), designated by a number, called the problem type. At this point, all seven questions have been asked and decision processes that threaten either decision quality or acceptance have been eliminated. What is left over is the *feasible set*.

For some problem types there is only one alternative remaining in the feasible set. However, the majority of feasible sets have more than one suitable decision process. The basis for choosing among alternatives in the feasible set lies upon a *time-efficient/time-investment continuum*. The time-efficient bias promotes the premise that more participative methods are slower than those that are less participative and use up more time on the part of each of those involved in the decision. Consequently, the most autocratic choice within the feasibility set is selected. At the other end of the continuum, the time-investment bias promotes the idea that participation has beneficial organisation-development consequences. The facilities managers should use their judgement to choose the appropriate point along this continuum for a given problem.

For example, when managing people through change, one of the most important tasks in which the facilities manager is involved at an early stage is assessing the level of participation to be offered to those people being impacted by the change (see Chapter 5 on Managing People Through Change for a fuller discussion of this issue). A worked example of a decision body flowchart analysis is shown later in Figure 8.9.

Inputs

Relevant information about the problem is required to complete the situational analysis of the problem. For example, when deciding upon the level of participation to be given to employees undergoing an office refitting, the facilities manager consulted the upper management of the client group, as well as drawing upon previous managing people through change experiences.

Output

- A problem that has been allocated an appropriate decision body.

8.2.6   Step 5: Establish a decision-making process plan

Step objective

- To produce and agree an outline plan for the decision-making process as a whole.

Step rationale and context

Planning is essential for good decision making. It provides the basis for integrating and coordinating all the tasks within the decision-making process. In particular, planning allows the manager to, firstly, collect and arrange relevant information, secondly, to identify and establish effective communication routes and, thirdly, to provide a programme of tasks that establishes intermediate and final deadlines that can be worked towards. For example, the reader will appreciate that the task to be carried out in a user-needs evaluation (see Table 3.2,

Table 8.3   Decision-making worksheet.

Decision-making worksheet
Problem description
Anticipated output
Assigned by   Date assigned Assigned to 1. Person responsible for output 2. Others involved/assigned
Key input information
Suggested sources of information
Further comments

Facilitation tasks) is basically a decision-making workplan: the *preparation* phase can be viewed as being synonymous with the exploration of the nature of the problem stage, while the *evaluation generic core* phase can be considered as being the same as the generation of possible solutions stage; finally, the *sort data and respond* phase can be perceived as being the choice among possible solution stages.

Tasks

- Identify key organisational resources required for the decision-making process.
- Establish a plan of work for the decision-making process.

Tools

Decision-making worksheet
The worksheet, shown in Table 8.3, allows the facilities manager to record the description of the problem, the anticipated outcome of the decision-making process, who is involved and key information requirements and sources.

Table 8.4   Decision-making process plan.

Decision-making process stages	Who	Man-days	Calendar date
Exploration of the nature of the problem			
Generation of alternative solutions			
Evaluation of alternative solutions			
Implementation of chosen solution			
Follow-up and control			
Referenced notations			
A:	D:	G:	
B:	E:	H:	
C:	F:	I:	

Decision-making process plan

Table 8.4 allows the facilities manager to plan easily how the decision-making process will be carried out – what needs to be done, who will do it and when it has to be done by. The implementation, follow-up and control stages have been included for completeness.

Inputs

Relevant information is required about the availability of organisational resources, such as people and time, to complete the decision-making process plan. For example, when formulating a disaster planning procedure for a large office block, the facilities manager carefully took into consideration the availability of short-term resources at any given time, so that a contingency element could be built in to allow for resource variations.

Output

- A plan that provides direction for the remaining stages of the decision-making process.

8.3 Stage 2: Generation of possible solutions

8.3.1 Introduction

The objective of this stage is to search for information that can be processed into a range of possible solutions. The emphasis of this stage is on effective and efficient information collection, and on creative, idea-generating techniques. Figure 8.5 shows the steps in the generation-of-possible-alternatives stage.

8.3.2 Step 1: Collection and analysis of information

Step objective

- To gain a better understanding of the problem context.

Step rationale and context

There is a tendency for managers to skip from the setting of initial objectives to the generation of possible solutions. However, this shortcut prevents the decision maker from gaining an improved understanding of the reasons for making the decision in the first place. The aim of this step is to address the problem within a wider context, providing a firm and balanced platform from which to generate possible solutions. This process involves two interacting activities: firstly, the collection of information around the problem area and, secondly, the analysis of this information.

Tasks

- Collect information related to the problem.
- Interpret information.

Tool

- Systematic fact finding and analysis technique.

This technique takes the form of a checklist, shown in Table 8.5, which aims to integrate the collection of information and its analysis. The purpose of the What, How, Where, Who and How Many questions is to identify all the significant

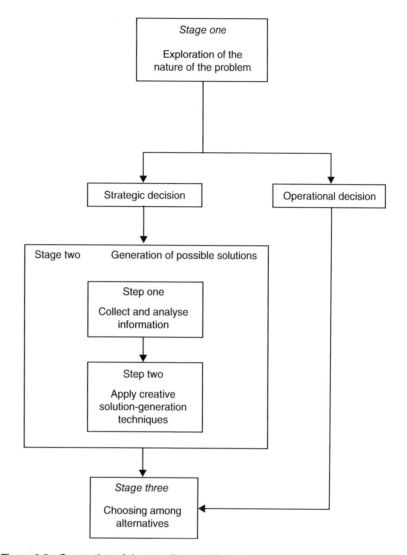

Figure 8.5   Generation of the possible solution stage.

issues that should be considered when generating possible solutions. The first Why? question is aimed at generating the major reasons why things are the way they are. The follow-up Why? question asks the decision maker why those major reasons have been put forward. A worked example is given later in Table 8.12.

Input

- Output from Stage 1: Exploration of the nature of the problem.

Output

- A problem statement placing the problem in context with its environment.

Table 8.5 Systematic fact finding and analysis checklist.

Systematic fact finding and analysis checklist		Date Problem description		
What activities are carried out?	⇒	*Why?*	⇒	*Why?*
How are things done?	⇒	*Why?*	⇒	*Why?*
Where are things done?	⇒	*Why?*	⇒	*Why?*
Who carries out activities?	⇒	*Why?*	⇒	*Why?*
How many times done?	⇒	*Why?*	⇒	*Why?*
Summary				

8.3.3 Step 2: Apply creative solution-generation techniques

Step objective

- To produce a range of creative possible solutions.

Step rationale and context

Ideally, managers would be able to generate all the possible alternatives in order to select the optimum solution. However, rather than optimise, managers generally follow a process of *compromising*, whereby they cease to generate possible solutions after they encounter the *first* alternative that meets some minimum standard of satisfaction. Such common managerial practice tends to limit severely the search for solutions. This is particularly undesirable for strategic decisions, where there are few or no precedents. The aim of this step, therefore, is to break out from these potentially rigid procedures and to infuse creativity techniques into the generation phase of the decision-making process. This has the effect of looking at problems from new angles, thereby improving the chances of generating better possible solutions.

It will be seen that the managerial tools discussed try to get away from the traditional 'round the table' meetings. Although such meetings set out to generate and refine ideas, in practice this rarely happens. Instead they often serve as a playing field for organisational politics, which are preoccupied with irrelevant issues or delegatable detail.

Task

- Apply creative techniques to generate innovative possible solutions.

Tools

SCAMPER attribute checklist

This technique, shown in Table 8.6, is essentially very simple and quick, and is perhaps most useful in situations where facilities managers want to develop basic ideas. The first stage for the facilities manager is to identify and pick out the major attributes of the problem under consideration. Each of these attributes in turn is then considered against a SCAMPER checklist of idea-spurring questions: Substitute? Combine? Adapt? Modify? Put to other uses? Eliminate? and Reverse?

Nominal group technique

The nominal group technique allows individual judgements effectively to be pooled within a group, and is particularly appropriate for strategic problems, where there is uncertainty about the nature of the problem and possible solutions. The ideal group size is 5–9, with sessions lasting a maximum of 60–90 minutes.

Part 1: Opening statement

The opening statement sets the tone for the whole session and should include the following components:

- the importance of the generation stage of the decision-making process should be noted;
- the group should be informed of the session's overall goal and how the results will be used;
- the four basic steps of nominal group technique should be briefly summarised (see Parts 2 to 5 below).

Part 2: Individual generation of ideas in writing

Ensure that all group members have a written copy of the problem question. The facilities manager should read the question aloud to the group and ask them to respond to it by writing their ideas in brief statements. This stage of proceedings should be done in silence and be around 4–8 minutes in duration.

Part 3: Round-robin recording of ideas

The facilities manager should explain that this phase is designed to map the group's thinking. The facilities manager should instruct one person at a time to present orally one idea from their list without discussion, elaboration or justification. This process continues until the group feels it has enough ideas.

Table 8.6 SCAMPER attribute checklist.

SCAMPER attribute checklist						Date	
Problem description							
Problem attribute	Substitute?	Combine?	Adapt?	Modify?	Put to other uses?	Eliminate?	Reverse?	Attribute summary

Group members should be encouraged to discuss others' ideas and to add new ideas, even though these items may not have been written down during Part 2. Ideally, the facilities manager should:

- record ideas as rapidly as possible;
- record ideas in the words used by the group member – the advantages of using the words of the group member are:
  - an increased feeling of equality and member importance,
  - a greater identification with the task,
  - the group feels its ideas are not being manipulated by the group recorder;
- record the ideas on a flipchart, numbering the items in sequence.

This stage provides an opportunity for all the group members to influence the group's decisions. The benefits of round-robin recording are:

- equal participation in the presentation of ideas;
- an increase in problem-mindedness;
- the separation of ideas from specific people;
- the increase in the ability to deal with a larger number of ideas – this is because people are more able to deal with a larger number of ideas if ideas are written down and displayed;
- a tolerance of conflicting ideas;
- the encouragement of 'hitchhiking' – hitchhiking is the notion that ideas listed on the flipchart by one member may stimulate another member to think of an idea not written down on the worksheet during the silent period;
- the provision of a written record and guide.

If facilities managers feel that the level of inhibition within the group may be such that it will discourage people to generate ideas, they may ask for the written lists to be passed in anonymously. The facilities manager then writes down one idea from each list in the same round-robin fashion.

Part 4: Serial discussion of the listed ideas

The facilities manager should explain to the group that the objective of this stage is to clarify the meaning and intent of the ideas presented. Each idea should be read out in sequence and comments should be invited. The members should discuss their thoughts on the importance, feasibility and merits of the ideas. Members may note their agreement or disagreement with an idea, but the facilitator should pacify arguments in order to save time. As soon as the logic of the idea is clear, the next idea should be discussed. The meaning of most of the ideas will be obvious to group members and discussion should be kept to the minimum. The facilities manager should allow around two minutes an idea.

The facilities manager should encourage the list of ideas as group property. Anyone can clarify or comment on any item. Within reason, new and/or amended items can be included, along with duplicated ideas being combined. However, the facilities manager should discourage the amalgamation of too many ideas into a single idea. Some group members may be trying to establish consensus through this process and the precision of the original ideas may be lost.

Part 5: Voting

Each group member should receive five cards. The members should select the five most important ideas and write one in the centre of each card. They should record the item's unique identity number in the upper left-hand corner. There should be a time limit of around five minutes for this stage and it should be completed in silence.

When all the group members have completed their five cards, the rank-ordering process should begin. The process should be as follows:

1. Each individual should spread their cards out in front of them so that they can see all the cards at once.
2. They should decide what they consider to be the most satisfactory solution to the problem from these five cards and write '5' on it. Turn the card over.
3. They should decide what they consider to be the least satisfactory solution to the problem from the remaining four cards, and write '1' on it. Turn the card over.
4. They should decide what they consider to be the most satisfactory solution to the problem from the remaining three cards and write '4' on it. Turn the card over.
5. They should decide what they consider to be the satisfactory solution to the problem from the remaining two cards and write '2' on it. Turn the card over.
6. The final card should be numbered '3'.

After this has been collected, the facilities manager should collect the cards and shuffle them together (to communicate to the group that no one is going to pay attention to how each individual voted). The vote should be recorded on a pre-prepared tally sheet in front of the group.

The facilities manager should then lead a discussion about the voting pattern. For example, if there is a polarisation between two extremely contrasting potential solutions, try to find out why. This helps the group achieve a sense of closure and accomplishment. If time permits, the group can further clarify the items and vote again.

Input

*   The problem statement from the 'collection and analysis of information' step.

Output

*   A range of possible solutions.

8.4 Stage 3: Choosing among possible solutions

8.4.1 Introduction

The objective of this stage is to evaluate possible solutions against predetermined criteria in order to arrive at an optimal solution. This requires, firstly, the identification of the evaluation criteria and, secondly, a comparison of the alternatives using the selected criteria. Figure 8.6 shows the steps in the choosing-among-possible-solutions stage.

8.4.2 Step 1: Identify the evaluation criteria

Step objective

- Identify the principal criteria with which the possible solutions will be compared.

Step rationale and context

It is important for the facilities manager to determine what criteria should be used to carry out the evaluation phase. The better a decision maker can distil the more important criteria from the less important ones, the better the final choice will be.

The objectives established in the 'exploration of the nature of the problem' stage provide the basis for selecting the criteria for evaluation. It is prudent, however, to reconsider these objectives at this stage, in view of the specific possible solutions being considered. The evaluation criteria should reflect:

- the *feasibility* of each solution;
- the *acceptability* of each solution;
- the *vulnerability* of each solution.

The *feasibility* of an alternative measures whether there are sufficient physical, human and financial resources available within the organisation to implement it successfully. The *acceptability* of an alternative is a measure of what return is likely from choosing that alternative. The final criterion, *vulnerability* of an option, indicates the level of risk associated with an alternative. These criteria elements, and their relationships with each other, are shown in Figure 8.7 and will be discussed in Steps 2 to 4 below.

Task

- Define evaluation criteria.

Tool

Evaluation criteria checklist
This checklist, shown in Table 8.7, aims to encourage the decision maker to reconsider the initial objectives in light of the information and understanding gleaned

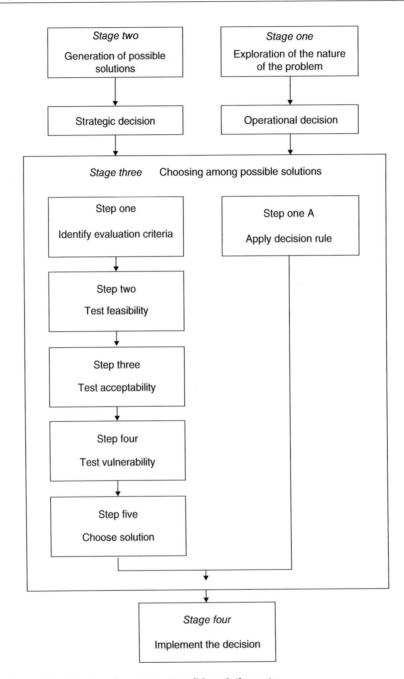

Figure 8.6 The choosing-among-possible-solutions stage.

from the decision-making process to this point. The facilities manager should record the initial objectives in the left-hand column and then question these objectives, with the question Why? in the middle column. Based upon this process of reflection, the objective should then be represented as an evaluation criterion

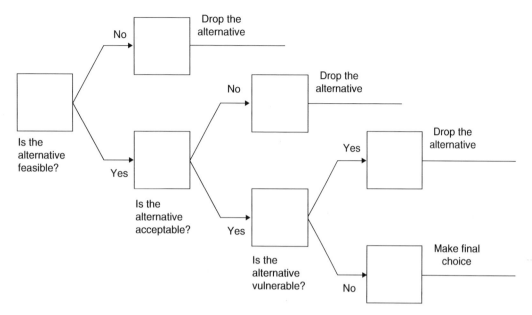

Figure 8.7   The evaluation process.

Table 8.7   Evaluation criteria checklist.

Evaluation **criteria checklist**	**Date** **Problem description**			
Initial objectives	$\Rightarrow$	**Why?**	$\Rightarrow$	**Evaluation** **criteria**
Summary				

in the right-hand column. The decision maker should read Steps 2 to 4 below before carrying out this procedure. A worked example is given later in Table 8.13.

Input

- Output from Stages 1 and 2.

Output

- A set of performance parameters against which possible solutions can be evaluated.

8.4.3 *Step 1A: Apply the decision rule*

Step objective

- To work through the decision rule to arrive at a solution.

Step rationale and context

It has been seen that operational problems are characterised by being routine in nature (see Stage 1: Step 3). Repetition of a decision situation provides both the opportunity and incentive to develop a decision rule that instructs the decision maker what solution to choose. All the decision maker has to do is feed the required information into the decision rule to determine the appropriate solution.

This process has the potential advantages of reducing the managerial input required in the decision-making process, as well as producing good-quality, consistent solutions. An obvious example where this 'automatic' decision-making process is desirable is in the event of a disaster. The facilities manager will want to develop decision rules for forecastable activities that will need to be carried out, leaving him to concentrate his efforts on unforeseen events that may occur.

Tasks

- Collect the required information for the decision rule.
- Work through the decision rule.

Tools

- Not applicable.

Input

- Problem characteristic (Stage 1: Step 3) and decision-rule input information.

Output

- A chosen solution ready for the implementations stage.

8.4.4   Step 2: Test feasibility

Step objective

- Assess availability of resources required for each possible solution.

Step rationale and context

The facilities manager, when evaluating the feasibility of a possible solution, should compare the required levels of resources required by the option with the actual resources available. If the resources required by a possible solution are not readily available, the proposed solution is not feasible. The three important elements when assessing the resource requirements of an option are: skills, capacity and the 'degree of fit'.

Every potential solution option will require a collection of *skills* to be present within the organisation, so that it can be successfully implemented. If a possible solution requires a course of action that is similar to other activities within the organisation, then it is likely that the necessary skills will already be present. If, however, the implementation of the solution will require a completely new set of activities, then it is necessary to identify the required skills and to match these against those existing in the organisation.

Similarly, each possible solution will need a *capacity* requirement of resources, such as finances, space, human resources and so on. These capacity requirements need to be determined to estimate the resources necessary to implement the solution.

Finally, possible solutions cannot be evaluated in isolation, but have to be assessed in context with existing organisational activities. The *degree of fit* of an option indicates the extent to which the organisational consequences of a solution being implemented are compatible with other organisational activities. This notion is particularly relevant in outsourcing (see Chapter 4), where it is agreed that the user's environment towards facilities management outsourcing is as important, if not more important, than purely economic arguments.

Tasks

- Determine the required technical or human skills that are required to implement the option.
- Determine the required capacity resources that are required to implement the option.
- Determine the option's '*degree of fit*'.

Tool

Feasibility checklist
A helpful way of evaluating the feasibility of a proposed solution is to consider it against the checklist shown in Table 8.8. A worked example is shown later in Table 8.14.

Table 8.8 Feasibility checklist.

Feasibility checklist Evaluation criterion	Date Proposed solution Response
Skills requirements	
Capacity requirements	
Degree of fit	
Summary	

Input

- Evaluation critera (Stage 3: Step 1) and range of possible solutions (Stage 3).

Output

- Two sets of possible solutions: one set consists of solutions that are not feasible and are thus rejected while the other set consists of the solutions that are feasible and are ready to be tested for their acceptability.

8.4.5 Step 3: Test acceptability

Step objective

- Assess the extent to which the possible solution satisfies the objectives.

Step rationale and context

The degree to which a possible solution complies with the decision objectives can be evaluated on its operational and financial impacts.

The assessment of the *operational impact* of each alternative should be based on the following:

1. *Technical specification*. Does the proposed solution increase the chance of the service or product that the operation generates being closer to what the internal/external client wants?
2. *Quality*. Does the proposed solution reduce the likelihood of errors occurring in the creation of services or products?

3. *Responsiveness.* Does the proposed solution shorten the time internal/ external clients have to wait for their services or products?

4. *Dependability.* Does the proposed solution give an increased chance of things occurring when they are supposed to occur?

5. *Flexibility.* Does the proposed solution increase the flexibility of the operation, either in terms of the range of things that can be achieved or the speed of changing what can be achieved?

Financial evaluation involves predicting and analysing the financial costs to which an option would commit the organisation and the financial benefit that might accrue from the decision.

Tasks

- Determine the operational impact of the option.
- Determine the financial impact of the option.

Tool

Acceptability checklist

A beneficial method of evaluating the acceptability of a proposed solution is to consider it against the checklist shown in Table 8.9. A worked example is shown later in Table 8.15.

Table 8.9   Acceptability checklist.

Acceptability checklist	Date Proposed solution
Evaluation criterion	**Response**
Operational impact • Technical specification • Quality • Responsiveness • Dependability • Flexibility	
Financial impact	
Summary	

Input

- Evaluation critera (Stage 3: Step 1) and range of possible solutions (Stage 3: Step 2).

Output

- Two sets of possible solutions. One set consists of solutions that are not acceptable and are thus rejected. The other set consists of the solutions that are acceptable and are ready to be tested for their vulnerability.

8.4.6   Step 4: Test vulnerability

Step objective

- Assess the level of risk associated with a possible solution.

Step rationale and context

The risk inherent in any option can be the result of the facilities manager's inability to predict any of the following:

- the internal effects of an option within the organisation;
- the environmental conditions prevailing after the decision is taken.

Although it is unrealistic to expect accurate predictions of such variables, it is helpful if the decision maker can assess the broad range of risk for a few critical evaluation factors.

Task

- Determine the risk inherent in an option.

Tools

Downside risk analysis
Perhaps a simple, but nonetheless powerful, method of assessing risk is to assess the worst possible outcome for a given evaluation factor for the possible solution. Once this has been stated, the facilities manager then asks the question: 'Would the organisation be prepared to accept such a consequence?' For example, even though the expected payoff of option B (shown in Figure 8.8), is greater than option A, the downside risk of option B might be too great a risk for the organisation to bear.

When considering risk, the facilities manager should be aware of the *risky shift phenomenon*, where groups tend to choose more risky options than would otherwise be chosen if the decision was being taken by an individual. Although the cause of this phenomenon is not fully understood, it is generally considered to be the product of:

- *Diffusion of responsibility*. When acting individually, decisions generally tend to be more heavily influenced by considerations of what happens if the

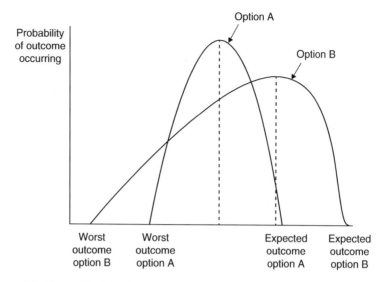

Figure 8.8   Downside risk distribution.

potential solution goes wrong. However, when individuals feel that the risk element is shared throughout the group, they will be more prone to accept more risk than would otherwise be acceptable.

- *Leadership of the group by high-risk takers.* There is a tendency for individuals who have a naturally high tolerance of risk to gravitate towards leadership roles within the group, thereby enabling them to be in a position to guide the group towards the more risky potential solutions.
- *Peer pressure.* It has been observed that individuals may feel under pressure to adopt higher-risk solutions than they would personally like, in response to a perceived stigma attached to being unduly cautious or conservative.

To counteract such forces the facilities manager may wish to adopt some of the following precautions:

- Assign one group member to be a 'devil's advocate' to promote constructive criticism within the group.
- Ensure that group leaders suppress their personal preferences at the beginning of the discussion.
- Periodically bring in people external to the group who should be encouraged to question the group's assumptions.
- Allow a follow-up meeting to allow individuals a period of reflection.

Input

- Evaluation critera (Stage 3: Step 1) and range of possible solutions (Stage 3: Step 3).

Output

- Two sets of possible solutions. One set consists of solutions that are too vulnerable and are thus rejected. The other set consists of the solutions that are all feasible, acceptable and exhibit tolerable risk. From this latter set the final choice is made.

8.4.7   Step 5: Choose a solution

Step objective

- To choose a solution for implementation.

Step rationale and context

Ideally, the choice step will result in the selection of the solution with the highest possible payoff. However, the facilities manager will no doubt be painfully aware that the evaluation of alternatives rarely produces one clear solution and ultimately the solution chosen will be based on individual judgement or group consensus. No matter what solution is chosen, the facilities manager should remember that it can often be made to work through effort and skill in its implementation.

Task

- Choose the option to be implemented through group consensus.

Tool

Modified nominal group technique
By modifying the nominal group technique, the facilities manager can pool individual judgements and work towards a consensus of what the chosen solution should be. The process should start off at Stage 3: Step 4: Serial discussion of the listed ideas, with the list of ideas being replaced by the short list of possible solutions that have passed through the evaluation steps. The voting phase will be aimed at making a final choice, rather than a short list of possible solutions.

Input

- Final output of evaluation process (Stage 3: Step 4).

Output

- A chosen solution ready for the implementations stage.

8.4.8   How to implement, follow up and control a decision

At this stage the facilities manager will have a solution that is ready for the implementation, follow-up and control stages. These final stages close the problem-solving cycle (see Section 8.1.4).

The *implementation* stage involves the required planning and carrying out of activities so that the chosen solution actually solves the problem. Insufficient managerial care with the implementation stage is a primary reason why good solutions often fail. In particular, managers should be wary of:

- the tendency not to involve those people impacted by the decision within the implementations stage, often leading to employee resistance to the decision (see Chapter 5 on Managing People Through Change);
- the tendency not to provide appropriate resources for the implementation phase – many implementations are less successful than they could have been because of inadequate resources, such as time, staff or information.

The *follow-up and control* stage involves the facilities manager making sure that what actually happens is what was intended to happen. To enable the smooth running of this stage, the facilities manager should set up the infrastructure in advance for the collection of the information necessary to monitor the implementation programme.

Case Study: Decision making

The decision-making process has been presented within this chapter as having three sequential stages: the exploration of the nature of the problem, the generation of possible solutions and the choosing among possible solutions. This case study aims to illustrate this sequence and points of interest within it.

Description of the organisation

This example of a decision-making process took place in a management company that had recently taken over management of a residential estate of some 500 dwellings from the original developer. The management company consists of six directors who are in charge of formulating policy on all aspects of the running of the estate. The company employs a managing agent whose role is to implement the directors' policy. All the work carried out on the estate (such as the maintenance and repair of the structure and fabric of the estate, along with the cleaning and upkeep of communal facilities such as gardens, car parking, intercoms, gymnasiums and laundry areas) is contracted out.

Initially, the management company were bogged down in developing administrative regulations and procedures. On reflection, many of the directors felt that many of their earlier decisions were rushed, ill-considered and reactionary in nature. The management company wanted to inject some sort of structure into their decision making, in the hope that better-quality decisions would be generated. Towards this aim, the directors used this book as a guide to good decision making, adapting and building upon it to form processes more suited to their organisation and culture.

Exploration of the nature of the problem

Sensing the problem

A problem gap between the actual and desired performance of the *day-to-day maintenance* function was sensed by the directors. The problem was identified through two channels. Firstly, it was observed that there was an upward pressure on day-to-day maintenance expenditure. This was not readily obvious as the budgeting categories combined planned and day-to-day maintenance. Secondly, there was an increase in the number of complaints from leaseholders about the speed and quality of repairs being carried out.

The first action taken by the management company was to set up a subcommittee to investigate how information was being processed and presented by the managing agent. The medium- to long-term aim was to develop a management information system that focused on critical areas and placed directors in a position where they could sense problems at the *opportunity* stage, rather than the *crisis* stage.

Setting initial objectives

The setting of initial objectives was undertaken by the directors using the SMART checklist. The directors felt that the checklist helped them focus on the critical elements of the problem, generating both direction and content for the remainder of the decision-making process. Previously, the directors tended to work towards extremely vague objectives, and were subsequently prone to 'managerial meandering'. Table 8.10 shows the management company's decision objective checklist.

Identifying problem characteristics

This stage of the decision-making process was perceived as being especially useful by the directors. In particular, it stimulated discussion about the trade-off between short-term operational decisions and medium- to long-term strategic decisions. The directors felt the main lesson learnt from this stage was the critical importance of injecting context into any solutions generated. Table 8.11 shows the management company's decision-type diagnostic checklist.

Establishing a decision-making group

The directors felt that the eventual decision-making group was intuitively obvious from the outset. However, they did feel the exercise was useful, in that it established the need to consult the leaseholders explicitly about the day-to-day maintenance policy.

Using the situational leadership analysis outlined in Section 8.2.5, the directors came up with the following answers, shown in Figure 8.9:

A. Is there a quality requirement such that one solution is likely to be preferable to another? *Yes.*

B. Do you have sufficient information to make a high-quality decision? *No.*

Table 8.10   Decision objective checklist: an example.

Decision objective checklist	
Problem description	*A gap between the actual and desired performance of the day-to-day maintenance function.*
Proposed objective	*Day-to-day maintenance should be: (1) of a high quality; (2) cost effective; (3) carried out within an appropriate time limit; (4) integrated with the planned maintenance policy; (5) easy to manage.*
Problem characteristics	**Comments**
Specific. Is the proposed objective sufficiently clear to avoid ambiguity and uncertainty?	*Although they are qualitative in nature at the moment, it is envisaged they are specific enough to be getting on with. It is noted that objectives (4) and (5) have ramifications for the planned maintenance, finance and administrative functions. These other areas will be addressed simultaneously, in the anticipation of creating synergy between the functions.*
Measurable. Does the proposed objective enable its performance to be evaluated?	*The bottom line cost of the repairs is envisaged as being measurable, along with the speed they are carried out. 'Hard' evaluation of objectives (1), (4) and (5) is seen as difficult. The quality issue in particular must be clarified within the eventual solution.*
Attainable. Is the proposed objective realistically attainable?	*Not withstanding the fact that the objectives are a bit vague at the moment, it is anticipated that the objectives are attainable.*
Relevant. Is the proposed objective consistent and linked to other organisational objectives and processes?	*All the objectives are extremely relevant to the core mission of the management company: to maintain and preferably enhance the asset value of the leaseholder's property.*
Trackable. Does the proposed objective enable progress towards its accomplishment to be monitored?	*It has been identified that there must be changes in the management information system if the progress of these objectives is to be monitored. Again, concern is expressed that the objectives (4) and (5) were difficult to monitor. Intuition to the fore!*

Table 8.11 Decision-type diagnostic checklist: an example.

Decision-type diagnostic checklist		
Problem description	*A gap between the actual and desired performance of the day-to-day maintenance function.*	
Problem characteristics	**Operational**	**Strategic**
Rarity. How frequently do	Not rare	Very rare
similar problems occur?	Notes: *This is the first time that a day-to-day maintenance policy is being formulated. It is hoped the exercise will not have to be repeated on a regular basis.*	
Radicality of consequences. How	Not radical	÷ Very radical
far is the solution of the problem likely to change things within the organisation?	Notes: *Day-to-day maintenance, and its interaction with planned maintenance, is the critical variable for the overall success of the estate. Good maintenance will enhance the estate, as well as keep a downward pressure on the leaseholders' service charge.*	
Seriousness of consequences.	Not serious	÷ Very serious
How serious would it be for the organisation if the chosen solution of the problem went wrong?	Notes: *See above.*	
Diffusion of consequences. How	Not widespread	÷ Very widespread
widespread are the effects of the decision likely to be?	Notes: *As highlighted in the initial objectives, it has been identified that the day-to-day maintenance policy will have ramifications for other managerial areas such as finance, administration and planned maintenance.*	
Duration of consequences. How	Not long	÷ Very long
long are the effects of any decision likely to remain?	Notes: *Adequate day-to-day maintenance, in conjunction with an effective planned maintenance programme, will reduce maintenance costs in the medium to long term.*	
Precursiveness. How far is the	Not precursive	÷ Very precursive

Table 8.11 (cont'd).

solution of the problem likely to set parameters of subsequent decisions?	Notes: *It is hoped that the formulation of an effective day-to-day maintenance policy will limit the range of problems in the future to those of a technical, rather than those of a managerial nature.*
Number of interests involved.	Few parties ÷ Many parties
How many parties, both internal and external to the organisation, are likely to be involved in the solution of the problem?	Notes: *It is envisaged that the final solution will be based on input from the directors, leaseholders and the contractors.*

Summary
The formulation of a day-to-day maintenance policy is definitely extremely important for the estate and is strategic in nature.

C. Is the problem operational in nature? *No.*
D. Is the acceptance of the decision by others critical to effective implementation? *No.* (Although a willing acceptance of the final solution by the managing agent would be useful, it is not essential.)
E. Do others share the organisation goals to be obtained in solving this problem? *No.* (The directors suspect that the managing agent and contractors will, in the short term, resist any solution that will make them more accountable for their actions.) It should be noted, therefore, that question G does not have to be asked in this particular case.

From this analysis, the resultant feasible set was C2. In other words, the directors would consult other parties, such as contractors and the managing agent. However, the final decision would be made by the directors alone, and may or may not reflect the others' opinions.

Establishing a decision-making plan
The management company drew up a plan, in particular highlighting critical milestones. The stage proved useful in maintaining a sense of tension throughout the decision-making process.

Generation of possible solutions

Collection and analysis of information
Information was gathered from two sources: the managing agent and the leaseholders. The directors viewed the former source as corresponding to the *actual* state of affairs, while the latter source indicated the *desired* state of affairs.

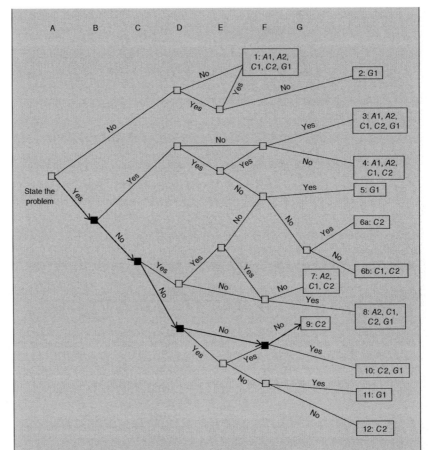

Figure 8.9   Situational leadership analysis flowchart.

The information collected from the managing agent fell into the following topics:

1. Specifications
2. Tendering arrangements
3. Contract supervision and control
4. Managing agent/leaseholder interface
5. Managing agent/management company interface
6. Interaction between planned and day-to-day maintenance

The information collected from the leaseholders fell into the following areas:

1. Leaseholder/managing agent interface
2. Perceived quality of maintenance carried out
3. Perceived priority of work that should be carried out

From this information a comprehensive picture could be developed. The directors felt that the systematic fact finding and analysis checklist was a useful way of breaking down and simplifying the analysis of the problem. It became clear that although the problem symptoms manifested themselves at the contractor and managing agent level, the actual cause was at a management company level. Table 8.12 shows the management company's systematic fact finding and analysis checklist.

Table 8.12   Systematic fact finding and analysis checklist: an example.

Systematic fact finding and analysis checklist	Data Problem description	Day-to-day maintenance
What **activities are carried out?** ⇒	*Why?* ⇒	*Why?*
A vast range of day-to-day repairs are carried out. Many of the repairs that have been carried out were on building elements that were due for replacement/refurbishment on the planned maintenance programme.	*Lack of integration between day-to-day maintenance and planned maintenance.*	*Lack of direction and policy from the management company.*
How are things done? ⇒	*Why?* ⇒	*Why?*
Generally speaking, the sequence of events is as follows: a leaseholder reports a problem, the managing agent comes out and inspects the defect, and if needs be contacts a regular contractor to carry out the required works. There are rarely any specifications, tendering arrangements or contract supervision. In addition, response times are often extremely slow.	*Lack of procedures and regulations laid down by the management company, allowing the managing agent free rein to carry out shoddy, expensive repairs.*	*Lack of direction and policy from the management company.*
Where **are things done?** ⇒	*Why?* ⇒	*Why?*
Not applicable.		
Who **carries out activities?** ⇒	*Why?* ⇒	*Why?*

(continued)

Table 8.12   (cont'd).

Contractors carry out all the repairs. No information is demanded from the contractors to show they are members of any trade associations, have sufficient insurance cover, financial background and so on.	*Lack of procedures and regulations laid down by the management company, allowing the managing agent to employ contractors on the basis of convenience, rather than on quality, cost-effectiveness and timeliness*	*Lack of direction and policy from the management company.*

How many times done?	⇒	*Why?*	⇒	*Why?*
Not applicable.				

Summary

The present problems are a result of a lack of planning, coordination and control of the managing agent by the management company.

Application of creative solution-generation techniques

In order to stimulate the generation of creative solutions, the nominal group technique was applied. The directors felt that previously meetings tended to revert to a general discussion about the nature of the problem, rather than how to solve it. Although the process of using the nominal group technique was felt to be 'silly' by all of the directors initially, on reflection there was a consensus of opinion that it was a useful way of reaching meaningful solutions quickly.

The process outlined earlier in this chapter was followed. The opening statement was defined as:

> What should be done in order to ensure a well targeted day-to-day maintenance which provides a good quality, cost effective and responsive service?

In this particular instance all the solutions put forward at the final voting stage were aspects of, or permutations of, a measured term contract (MTC).

Choosing among possible solutions

Identifying evaluation criteria

This stage was separated from the generation phase, in an explicit effort to allow reflection, and not to get 'carried away' and rush headlong into a decision. Firstly, the generation phase was reviewed and the notion of an MTC was clarified so that all the directors had a consistent understanding of the possible solution to be evaluated. It was agreed that an MTC was one in which contractors tendered for the right to carry out all work of a certain type for an agreed period.

The directors moved on to developing the evaluation criteria. Many of the directors felt that they had covered similar ground when they set the initial objectives. However, they did feel that the exercise was a suitable moment to reflect on the decision-making process to date and to make sure that the decision-making process was on the right path for the remaining stages. Table 8.13 shows the management company's evaluation criteria checklist.

Table 8.13   Evaluation criteria checklist: an example.

Evaluation criteria checklist	Data Problem description	Day-to-day maintenance
Initial objectives ⇒	**Why?** ⇒	**Evaluation criteria**
Quality	Good quality throughout the repair process is essential.	The solution should promote good quality at all stages of the repair process.
Cost-effectiveness	Cost-effectiveness is essential if a downward pressure on service charges is to be maintained.	The chosen solution should maximise competition, but not at the expense of compromising service quality.
Responsiveness	Repairs should be carried out as quickly as possible in order to limit the amount of damage and to satisfy the leaseholders.	The chosen solution should enable an appropriately quick and reliable repair service.
Integration with planned maintenance	Through the integration of the planned and day-to-day maintenance programmes, maintenance costs can be reduced and the repair service better targeted.	The chosen solution should enable repair activities to be accurately targeted to create synergy with the planned maintenance programme.

(continued)

Table 8.13 *(cont'd).*

Manageability	The policy should allow the management company to monitor the performance of both the managing agent and the contractor.	The chosen solution should be easy to organise and control by the management company.

Summary *The chosen solution should be predominantly geared towards quality of service and ease of management company control, rather than bottom line cost.*

Table 8.14 Feasibility checklist: an example.

Feasibility checklist	Date Proposed solution *Measured term contract*
Evaluation criterion	**Response**
Skills requirements	*Management company. The MTC would facilitate easier and more consistent control for the directors. There is confidence that no new skills were required.*
	Managing agent. Concern is expressed whether they have sufficient skills to carry out their envisaged new tasks. On balance, however, it is decided that it is more important to fit the managing agent to the task, rather than the other way round. The situation will be closely monitored and if the performance of the managing agent is found to be lacking, a replacement will be found.
	Contractor. Concern is expressed whether one contractor will have sufficient expertise for the large number of building trades required. However, taking into account that all specialist maintenance functions are carried out under separate service contracts, there is sufficient confidence that one contractor will be capable of carrying out the basic repairs required of him.
Capacity requirements	*It is felt that there are sufficient resources to develop the MTC. In fact, even if additional resources are required in the short term, it is felt that they will be quickly recouped in the medium term.*
Degree of fit	*It is felt that the MTC would considerably contribute to the overall effectiveness of the running of the estate and could be easily integrated with other functions, in particular planned maintenance.*

Testing feasibility, acceptability and vulnerability

Once the evaluation criteria had been established, the directors then tested the feasibility of introducing an MTC. The directors were somewhat concerned that they were only evaluating one possible solution. To partially address this shortcoming, they split up into two groups, one half arguing for the solution, while the other half played the devil's advocate and argued against it. This arrangement was repeated for testing both acceptability and vulnerability. Tables 8.14 and 8.15 show the management company's feasibility and acceptability checklists respectively.

When testing vulnerability, the directors felt that the main risk involved was being locked into a contract with a contractor for one year. However, they felt the risk was acceptable for two reasons: firstly, there would be break

Table 8.15   Acceptability checklist: an example.

Acceptability checklist	Date Proposed solution *Measured term contract*
Evaluation criterion	**Response**
Operational impact	
• Technical specification	*By having an MTC, a detailed specification can be drawn up of likely work items. This will form part of the initial contract documentation and therefore will be enforceable.*
• Quality	*The specification can clearly define expected standards of material and workmanship, with reference to appropriate British Standards and Codes of Practice.*
• Responsiveness	*Work items within the Schedule of Rates can be prioritised in response times, e.g. 24 hours, 48 hours, one week and so on.*
• Dependability	*The terms of the MTC will stimulate the contractor to be dependable, to avoid the risk of liquidated damages.*
• Flexibility	*The MTC will tend to 'lock' in a contractor for the duration of the contract. It is appreciated that this cannot be avoided. On the plus side, however, a long-term relationship may be developed in both the client and the contractor's interests.*
Financial impact	*The financial implications of an MTC can be grouped into three main areas. Firstly, it enables more accurate cash flow predictions. Secondly, it promotes lower unit costs. Thirdly, in the medium to long term (if the repairs can be better targeted and be of an appropriate quality), savings can be made.*

clauses in the contract if gross incompetence and so on was being displayed; secondly, they felt that the tendering arrangements would tend to filter out potentially troublesome contractors, with the tender requesting references, details of past/current contracts, previous years' audited accounts, insurance arrangements, health and safety policy, and so on.

The final choice to begin developing an MTC was not taken immediately after the evaluation, but was taken at the following meeting. Again, this was an explicit effort to allow a period of reflection.

8.5 Conclusions

In conclusion, one of the directors summed up the use of combining theory, experience and judgement when making decisions when he said:

> I can't see us using these techniques all the time. I thought some of them were a bit contrived to say the least. But saying that (although I'm not sure if it was the techniques themselves, or the fact we were just made to think about what we were doing generally), at the end of the day we seemed to have produced better policies. I think that we will eventually drop them in their present form, but translate the concepts behind them into more tailor-made decision making procedures.

Therefore, it is mainly the thinking behind the stages and techniques that should persist to sharpen the facilities manager's approach to decision making. Often, of course, what is involved is not a journey to a single choice or something about which everyone will feel happy. Maybe it is some general concern that things could be done better, say more rapidly, but a low consensus about what range of issues this might involve and a suspicion that in fact a web of interlocking actions is needed to make progress. In this sort of more turbulent situation Stages 1 to 3 of the decision-making process model underpinning this chapter are especially important. These would involve a joint exploration with stakeholders of the nature of the problem, the generation of possible solutions and some choice between these alternatives. If a strong consensus can be achieved with stakeholders around these stages then this can be a strong basis for a campaign on a broad front in order to progressively move towards an improvement in the problem situation.

The following case study provides an example of a practical approach that can be used in this rather amorphous sort of situation. It is based on Kurt Lewin's (1947) notion of social (or organisational) situations being the result of 'quasi-stationary equilibria', i.e. the outcome of a fight between driving forces and restraining forces, with the current situation being the result of the balance between these at the given point in time. This is a very dynamic view of problem situations and leads directly to some simple but effective ways of supporting a joint problem-solving effort by stakeholders, in which a lot of the effort is concerned with first surfacing the relevant current issues, then gauging their strengths and then creatively seeking effective actions within this context.

Case Study: Decision making in a turbulent multi-dimensional situation

The example given here draws from an initiative to address the perceived underperformance of the construction industry. This was carried out via workshops with diverse senior stakeholders in five countries. It has been described in more detail elsewhere as an element of the revaluing construction initiative that was broadly concerned (as is this book) with bringing to the forefront the issue of value for clients and users and orientating efforts to deliver this (Barrett, 2007, 2008). This example will draw on the work of the UK stakeholder group.

Lewin's forcefield model

The approach taken is underpinned by the notion of a forcefield analysis. The framework used for this is given in Figure 8.10.

It can be seen that for this example the scale to the left is labelled to reflect increased or reduced 'value' around the equilibrium of the value generated by the 'current approach'. Visual space is provided for a number of driving and restraining forces that have the capacity to increase or reduce value. For each force the option is provided to roughly scale before and after the situation.

Programme of work

Within this broad framework, the challenge is to engage stakeholders to identify and manipulate the main forces at play. For the revaluing construction initiative these were a diverse group of 24 senior construction participants. These were brought together by invitation for one day and worked within the programme set out in Figure 8.11. To address the knotty issue at hand various dimensions were represented in the programme. Firstly, the participants were briefed so that there was a clear, shared goal for the day and some common basis of knowledge to complement their individual expertise. The room used was set out in banquet style to facilitate a mix of plenary and group discussion sessions. For the first group session individuals were asked to work within a group of their stakeholder peers (clients with clients, designers with designers, etc.), which let easily into discussions about drivers and barriers to increasing value from construction. Of course, often these concerned other stakeholders, especially in the case of barriers! The groups were asked to brainstorm drivers and barriers and then prioritise into the most influential four of each. The further sifting occurred through a plenary feedback session within which ideas from each group could be pooled, discussed and refined. Thus, by

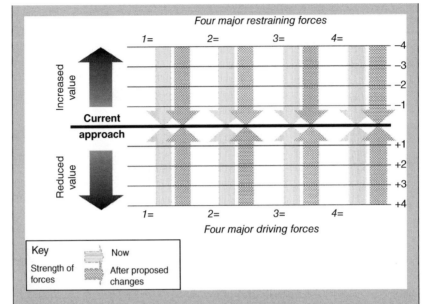

Figure 8.10   Lewin's forcefield analysis applied to revaluing construction (Barrett, 2008).

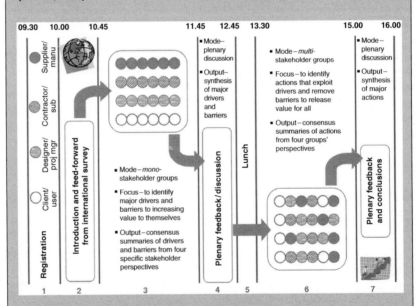

Figure 8.11   Programme for a revaluing construction workshop.

lunchtime a fairly long list of drivers and restrainers had been captured and their existing strength calibrated.

During the lunch break the workshop organisers summarised the forces on to a sheet, with a space in the middle for actions to be noted. This is shown in Figure 8.12.

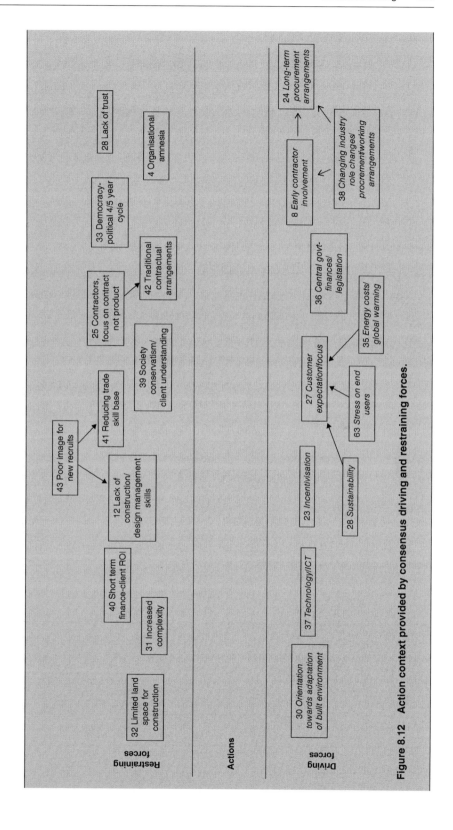

Figure 8.12   Action context provided by consensus driving and restraining forces.

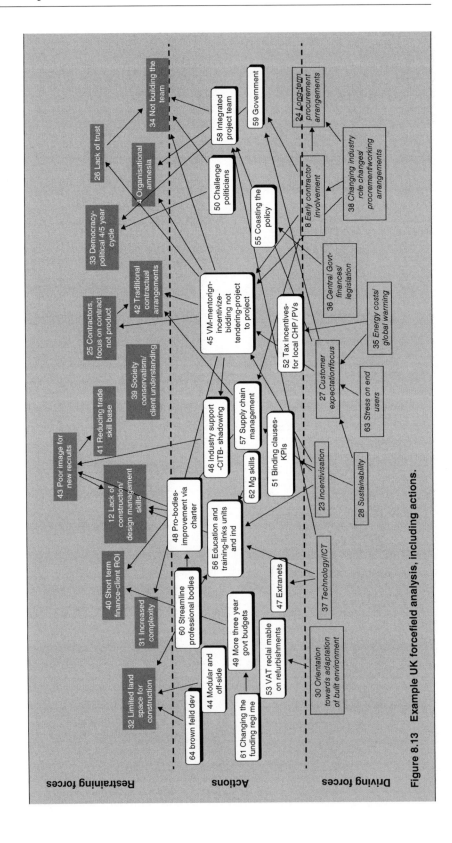

Figure 8.13 Example UK forcefield analysis, including actions.

In the afternoon the group work recommenced with a focus on finding actions that could be effective in practice. A significant difference was injected into the groups by reconfiguring them as multi-disciplinary problem-solving groups, where the various stakeholders now worked together to invent, explore and prioritise holistic solutions. The forces either side of the action space were used as a stimulus by asking in each case what action could increase the driving force or, in the case of the restraining force, bring about a reduction? This provided a starting point for the generation of many possible practical actions, which then were priorities in each group by their potential impact. Although actions usually surfaced from consideration of one force, those prioritised were then considered in terms of all of the drivers and restrainers they could be connected to.

As before, the work of the groups was collected through plenary discussion and jointly pooled, discussed, refined and summarised in the last session. The result was a forcefield made up of the main consensus driving and restraining forces, with the driving forces linked to, and energising, actions, which in turn connected to and 'attacked' restraining forces. The notion was that this provided a view of how this group felt the equilibrium line could be shifted upwards through a range of explicitly interconnected actions. The summary for the UK workshop is given in Figure 8.13 as an example.

This was in fact synthesised with results from four other countries and an international agenda for change created. This involved identifying major action themes (which connected to multiple forces) and the stakeholders that were empowered to break the deadlock of the status quo.

Although used at an international level here, this approach has been used productively within an organisation to address health and safety issues. This approach could, of course, be used by an FM unit to engage with the users of the spaces being managed. It could readily be extended to wider stakeholder groups. It can provide a vehicle to get people working together collectively, towards consensus ways forward.

References

Barrett, P.S. (2007). Revaluing construction: an holistic model. *Building Research and Information*, **35**(3), 268–286.

Barrett, P.S. (2008) *Revaluing Construction*. Blackwell Publishing, Oxford.

Elbing, A.O. (1970) *Behavioral Decisions in Organizations*. Scott, Glenview. p. 14.

Lewin, K. (1947) Frontiers in group dynamics. *Human Relations*, **1**(1), 5–41.

Vroom, V.H. and Jago, A.G. (1988) *The New Leadership: Managing Participation in Organizations*. Prentice-Hall, Englewood Cliffs, NJ.

IV Conclusions

9 Sustaining the Pursuit of Facilities Management Excellence

9.1 Introduction

9.1.1 Context

The previous chapters in this book have discussed the various systems that would be included in the 'ideal' facilities organisation (Chapter 1) and offered many options for facilities managers to consider. This chapter steps back and suggests ways in which performance can progressively be enhanced over the medium term.

In practice, few organisations have the opportunity to design a facilities organisation with a 'blank sheet of paper'. For most organisations there will be existing activities, probably fragmented, possibly with gaps in the provision and maybe with overlaps and some confusion. It is from this base that those seeking change must move.

Obviously, it would be theoretically possible to scrap the current facilities provision and start again, but this is not very realistic. Most organisations will seek to *evolve* a better provision from their existing base. Thus the question is how can an organisation consistently and continuously improve its facilities provision? Or where should it start and how can it keep going? This section seeks to address these questions.

9.1.2 Overview

It is quite easy to improve something specific, say to negotiate better terms on a contracted-out service for grounds maintenance. What is notoriously difficult is, consistently and in an integrated way, to achieve an ever better level of service over the medium to long term. What is needed for this is to take action simultaneously on a number of fronts. The facilities management function must have information on its performance on the full range of interactions; it needs to develop a broad strategic thrust to align individual actions with the core business strategy and with each other; it needs to develop the capabilities of staff so that they actively overcome problems and spot opportunities for doing things in better ways.

Facilities Management: The Dynamics of Excellence, Third Edition. Peter Barrett and Edward Finch.
© 2014 John Wiley & Sons, Ltd. Published 2014 by John Wiley & Sons, Ltd.

The objective, therefore, is to know what needs to be done, to know how to mesh individual actions to greatest effect and to have people involved who, once they know what is needed, are capable of, and motivated to produce, innovative solutions.

Although ultimately all of the above factors are needed for optimum performance it is more realistic to think in terms of a sequential development moving from a sound base of integrated feedback data to a clear strategic view with the capability of staff being enhanced by involvement in these activities and then through targeted efforts over the long term.

9.1.3   Summary of the different sections

- Section 9.1. Introduction.
- Section 9.2. A brief review of the nature of professional services is provided together with the implications for facilities management.
- Section 9.3. The use of multiple feedback mechanisms is illustrated as the primary means of first stimulating improvements and then sustaining the momentum created. An approach termed 'supple systems' is described.
- Section 9.4. The important area of strategic management is considered and advice given on how to create a longer-term orientation.
- Section 9.5. People are linked to the above plans and information in this section. The necessity of developing 'learning individuals' to support an improving organisation is illustrated.
- Section 9.6. Links back to the start of the book and then stresses that a holistic approach is essential to achieve a sustainable approach to facilities management excellence.

9.2   Client perception of facilities management services

9.2.1   Context

By definition those providing the facilities function are rendering a *service* to the core business. Given the objective of continuously enhancing facilities performance, it is important to appreciate fully the nature of services. The following section is designed to put the familiar aspects of facilities provision in a broader context. This context includes the perceptions of key *clients* of the facilities provision.

9.2.2   The nature of professional service quality

Intangibility of services

Facilities management is geared towards providing a service and hence its contribution to an organisation may be difficult to identify in concrete terms; there is no end product that can be held up and shown to the customer. The

Figure 9.1   Expectation/perception gap.

implications of this intangibility can be far reaching, especially in terms of the client's assessment of the facilities department's performance. The assessment of facilities services is likely to revolve around the client's *perception* of the service received compared with the client's *expectation* of the service (Gronroos, 1984). Thus the facilities group can make efforts in two distinctly different areas – namely managing the client's initial expectation and managing the client's perception of the service rendered, as shown in Figure 9.1.

It can be seen that *service quality* figures prominently in the above way of looking at performance from the client's point of view. Insights from the field of quality management will therefore be drawn upon at various points in this chapter.

Expectation

The client's expectation of any service will be conditioned to a great extent by past experience, but also by the initial messages concerning the service. Thus, the facilities manager should be careful not to overstate what the facilities department is capable of delivering. If this occurs, it is obvious that the client is unlikely to be satisfied with the service provided. However, in a similar vein, it is important to portray a positive, rather than negative, image.

For example, if an already overworked facilities department was approached about a possible office reorganisation, they should not just dismiss the proposal out of hand. Instead, after checking that the work was not urgent, they could take the initiative by saying that they would like the time to prepare reasoned proposals and then suggest a suitable date to discuss their ideas. In this way the message is conveyed that an in-depth analysis will be provided. It also demonstrates that the facilities department actively controls its time, and when a meeting is held, and an impressive presentation made, it will also confirm that the facilities group produces high-quality work, on time. This can be contrasted with an attempt to react immediately, in which case the facilities department would not provide itself with sufficient time to prepare a sensible solution.

Perception

The second area where effort can be directed is at the client's perception of the service rendered. It is important to stress that there is never any objective measurement of a professional service; it will always depend on individual assessments, which in turn will be based on multiple impressions from a variety of sources. It is, perhaps, easiest to highlight the problem by looking at one extreme situation: the job where everything goes right.

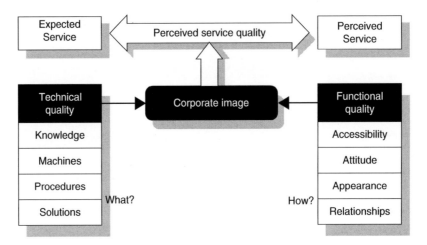

Figure 9.2   Technical and functional quality.

In such a case, the objective quality of the service provided is clearly high; however, it is quite possible that the client will be unaware of the facilities department's efforts simply because nothing has gone wrong. Doubtless the facilities group will have foreseen potential difficulties and taken action early on so that matters do not get out of control. Quite simply, they have avoided problems and, as a result, their work may appear deceptively simple to the rest of the organisation.

A positive reaction to this is for the facilities department to keep appropriate people informed of potential problems and the steps being taken so that it is clear that the input of the facilities group is positive, is proactive and is working. This approach involves taking the client through the job with you, rather than just handing over the final product. However, this suggestion should not be taken to an extreme and it would probably be inadvisable to provide the client with details of every decision, but the general suggestion of *taking the client with you* is very sound.

Technical and functional quality

Figure 9.2 builds upon Figure 9.1, but includes information on how the judgement between expectation and perception is conditioned. Two main categories of factors are shown, which together form the corporate image of the facilities department, which in turn provides the main input to the client's assessment of the service provided. The two main areas are concerned with *technical quality* and *functional quality*. The importance of the individual factors will vary for different organisations, but any assessment is likely to be a rich mix of both types of factors.

Technical quality is concerned with *what* is done and includes how well the problems were solved and the systems and techniques used. This is the area that facilities managers would normally be mainly concerned with.

However, facilities managers may be less inclined to consider the other factors, namely functional quality. The functional factors revolve around *how* the service was rendered. This includes items such as the appearance of staff, their attitude towards clients and how accessible and responsive the facilities department was to the client.

There is a growing body of research indicating that when clients judge the quality of a service, they give unexpectedly high weightings to the functional factors as well as the technical factors. It is, therefore, important for the facilities department to think through how they deal with the core business as an important, and quite separate, issue from what they do to solve the technical problems that they are faced with. (Chapter 5 illustrates how the facilities department can successfully manage people through change.)

Summary

It has been suggested that it is of critical importance to take the client with you in the delivery of the facilities function. This is important in really finding out what is required and implicit in giving consideration to how the service is rendered.

It can be seen that a key variable to be measured is the client's perception of facilities service quality. This is not to suggest that a purely reactive approach should be taken. Informing and explaining to clients the implications of various alternatives, often revealing conflicts between short-term and long-term solutions, is vitally important. It is, however, *not* advisable to say 'the client does not understand what they need, we will sort it out for – or possibly despite – them'.

We have heard this sentiment on many occasions and the following sections endeavour to demonstrate how a client orientation can be introduced, so that the technical skills present can be developed and applied to greatest effect.

9.3   Stimulating and sustaining improvements

9.3.1   General approach

The previous section highlights the fact that the assessment of facilities management services is likely to be highly subjective and consequently each client could well apply different criteria. From this it naturally follows that facilities management departments must interact with their clients if they wish to improve their services. Many facilities managers may argue that they already interact constantly with their clients – if they didn't they would not have any work to do. However, as the case studies in the previous chapter demonstrated, the purpose of this interaction is often purely to obtain instructions for the next task and as such is unlikely to lead to improved client services.

The key to achieving improved services is to introduce feedback mechanisms so that the facilities department actually learns from occasions where it could have done better or from opportunities/ideas from other sources. The range of possible sources of feedback/stimuli is shown on the generic model in Chapter 1 (Figure 1.1) and comprises:

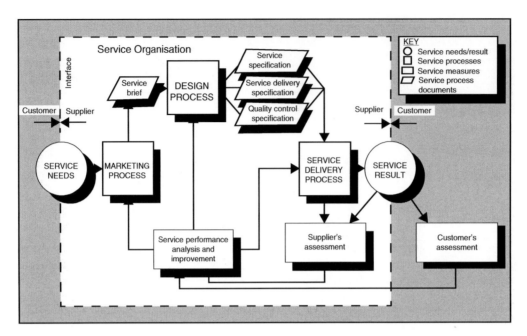

Figure 9.3   Service quality loop.

- interaction within the department between the facilities manager and the functional units;
- interaction between the facilities department and the core business;
- benchmarking of internal facilities services against other facilities organisations;
- interaction with the core to ascertain what future changes may occur to the business;
- scanning for possible developments within the facilities arena;
- interaction between strategic and operational facilities management to achieve a balance between current operations and future needs.

In simple terms, the facilities department continuously collects feedback externally, as well as internally, and then analyses the information and integrates it to achieve improved service levels. It should be stressed that the process is iterative and thus encourages continuous improvement. This emphasis was apparent in the International Organization for Standardization's Quality Management Loop for Services given in ISO 9004-2 (ISO, 1991) and is shown in Figure 9.3. The feedback mechanisms occupy the lower half of the diagram. This standard has since been superseded, but the current more generic ISO 9004:2009 (Managing for the sustained success of an organization – A quality management approach) maintains an emphasis on self-assessment, strategy and innovation.

Linking this diagram with the above listing of potential sources of feedback and acknowledging that it would not be sensible, or possible, to wait until the

Table 9.1   Key features of supple systems.

Feature	Comment
Client orientated	Above all the systems are tested against client requirements by actively seeking feedback through both hard and soft data.
Minimalist/holistic	'As much as you must, as little as you may', that is, not having systems for their own sake, but rather targeting high risk/gain areas. Better to have made some progress on all important fronts than to have a patchy provision.
Loose-jointed	The systems operate at an audit level: clarifying objectives, checking performance and integrating efforts. At an operational level different styles and approaches can be accommodated, especially when they have proved themselves over time.
Evolutionary	Allow incremental and continuing progress to be made from whatever base.
Symbiotic with social systems	Build on the norms and culture of the organisation, for instance allowing self-control or group pressure to operate where appropriate.

end of the provision of a service before getting some feedback, a richer picture emerges within the same general framework.

Based on this and aiming for systems that are client-responsive and facilitate a continuing cycle of improvements, an approach has been developed, through debate and observation, that endeavours to meet these criteria. The approach has been styled *supple systems* and is discussed next.

9.3.2   Supple systems

The key features of supple systems are given in Table 9.1 (Barrett, 1994). The following paragraphs consider each aspect in some more detail. However, in summary, the approach advocates that a strong, but flexible *audit* system is developed, which ensures that improvements in the quality of the service are being achieved. The audit system identifies sources of feedback, assesses if action is required, and at what level, prioritises between alternatives, allocates responsibility, checks later that action was taken, tries to objectively assess the impact of the actions and finally feeds these findings back to those involved.

Client orientated

The supple systems approach is in line with the ISO 9004-2 approach (Figure 9.3), in that emphasis is placed on the importance of interaction with the client. Thus, it can be seen that the above approach argues that feedback mechanisms should be introduced (or improved), so that existing systems can be tested to see if they really are meeting client requirements. It is important to remember that the quality of services will be assessed from a number of different perspectives, as discussed in Section 9.2, and consequently the facilities department should collect both hard and soft data. In this way true client orientation can be achieved and then improvements can be made from a basis of knowledge in such a way that maximum impact is achieved from the effort and energy available. Chapter 3 gives detailed advice on the crucial issue of measuring users' needs and

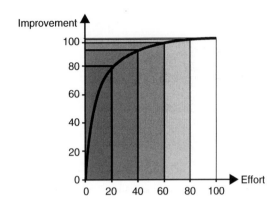

Figure 9.4   The 80/20 Rule.

perceptions. By emphasising feedback *first* the danger of fossilising defective aspects of systems is avoided.

Minimalist/holistic

'As much as you must, as little as you may' is a statement that should be applied to quality management systems, especially formal ones, and particularly when they involve the creation of paperwork. Given that resources are bound to be limited, it is important to consider how best to use the time and effort available. The well-known 'Pareto effect' or '80/20 rule' suggests that action in any one area will tend to be subject to diminishing returns (see Figure 9.4).

In practice this means that it is likely to be more effective to take some action on *all* major fronts rather than put all of the available effort in one direction. The generic model at the end of Chapter 1 (Figure 1.1) indicates the range of interactions. The idea being stressed here is that it is better to make some improvements on all dimensions rather than, say, putting all the available effort into creating brilliant contracts for contracted-out services, but being unsure of the contribution made to the core business strategy or the level of satisfaction of users with the facilities management provision, etc. This general approach also applies to the suggested sequencing of activities given in Section 9.1.2. Perfection should not be sought in one area before addressing the next.

This perspective casts doubt on the likely effectiveness of the traditional quality management approach, where importance is placed on the production of a 'quality manual'. Hence, organisational effort is concentrated on recording the systems so that they can be included in the manual, rather than considering where improvements could be made. Positively, ISO 9000 2000 is somewhat more client and improvement orientated.

Loose-jointed

One of the key aspects distinguishing the supple systems approach is the emphasis on an audit level of activity. This is in contrast with the usual quality

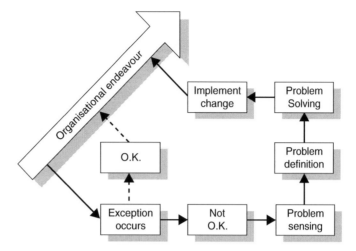

Figure 9.5   The process of planned change.

management approach which, as mentioned above, stresses recording all existing systems. The supple system method advocates that the existing systems can be left largely as they are, except when the feedback gained identifies a need for action. The audit system is designed to ensure that problem areas are spotted, the root cause identified, effective action taken and the lessons learnt, encoded in the formal system if appropriate. Consistency of effort towards satisfying core business needs *is* essential. Consistency at all levels in how this is done is not essential. This facilitates an organisation that can respond to individual needs using individual skills. Most importantly, it means time is not wasted scrapping and redesigning the things that are already done well. Thus effort is concentrated where it is needed.

Evolutionary

Where traditional quality management systems aim to produce consistency, i.e. manage the steady state, supple systems stress the management of the dynamic aspects of the firm. A continuous thrust of improvement is emphasised. As such, it is an approach that any facilities management organisation can take, from whatever baseline, keeping what already works well and flexibly addressing any problem areas identified. The essence of this is given in Figure 9.5 (Kast and Rosenzweig, 1985).

This suggested approach means that firms can be more certain that their performance is improving, without the heavy load of formally demonstrating that their existing systems (not performance) are all in order up to a minimum level at a given point in time. Thus over time the supple systems approach will lead to high performance, not just an adequate level but potentially well beyond. At the same time formal systems will have been created, but only where appropriate, generated by the feedback/improvement cycle given in Figure 9.5. This does not

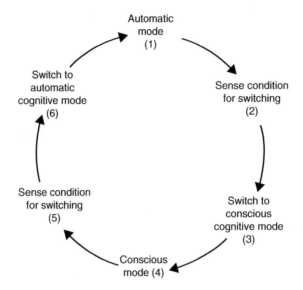

Figure 9.6 Switching cognitive gears.

mean the paper systems will grow endlessly. Where the feedback shows formal procedures are causing problems, they will be discontinued.

Thus, it will be possible to proceduralise steady-state activities for efficiency and reliability whilst still retaining a capability to identify key situations where change is demanded in order to ensure effectiveness and responsiveness. In short, the organisation will become adept at 'switching cognitive gears', as illustrated in Figure 9.6 (Louis and Sutton, 1991).

In this figure 'automatic mode' equates to proceduralised activities and 'conscious mode' to active problem solving. It is stressed that the real problem is knowing *when* to switch from one to the other. The feedback mechanisms are a key way of increasing awareness of when to move from automatic to conscious mode. Further advice on this is given in Chapter 8 in Table 8.2, where the criteria are set out that identify key dimensions of problems that need in-depth analysis. Moving from conscious to automatic mode, i.e. taking a specific solution and deciding to generalise from it using policies and procedures, can also be problematic. This is a key way in which experience from one part of the facilities organisation can be communicated to other parts. Solutions depend principally on judgement and awareness of the issue. Hard (paper) and soft (say, seminars) approaches should both be used.

Symbiotic with social systems

There is often a strong culture that itself can encourage the achievement of high-quality work. This can be an ethos of professionalism derived from pride in the discipline of facilities management. It can be something that has developed over time within the given firm, possibly generated by the example of key leaders in the firm. Whatever the foundation of the organisation's culture, every organisation

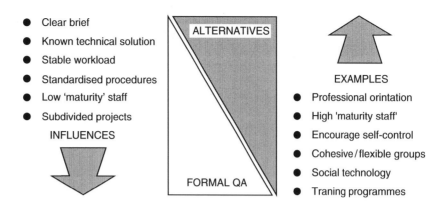

Figure 9.7   Factors supporting formal QA/alternatives.

is likely to have a distinctive set of norms. Generally, it seems reasonable to suggest the culture will have developed as a response to the particular nature of the organisation's workload as it interacts dynamically with the specific people making up the organisation's staff. The culture will not be an accident; it may even be appropriate to the circumstances!

Proposals for change should therefore take this existing aspect of a given organisation into account. More particularly, the development of supple management systems should aim, where possible, to work with the social systems in the firm so that the maximum synergistic effect $(2 + 2 = 5)$ can be achieved. This requires a systemic perspective which recognises that the formal and informal parts of the organisation may in some circumstances be considered as alternatives. This notion is illustrated in Figure 9.7 (Barrett, 1989). It can be seen that many of the alternatives are factors typically found in professional organisations. By acknowledging these factors the need for heavy formal systems can be kept to a minimum. This can be especially important in a small facilities department where informal mechanisms rightly predominate.

Summary

Overall, supple systems are proposed as an approach that can provide positive benefit to facilities departments by allowing any department, however small, to do something, start somewhere, in what will become a continuous development in the capability to identify and satisfy client requirements. As time goes on a robust, but flexible, framework of systems and processes will be created. At the same time staff should develop an understanding of their clients and of the key interrelationships within the organisation, leading to a successful service-orientated culture pervading the department and ensuring the effective and efficient delivery of facilities services.

For all but the smallest department a *senior* group meeting as a committee on a regular cycle can help ensure continued implementation over the longer term. It becomes a focus for collecting feedback and then prioritising effort.

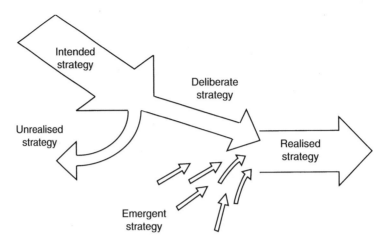

Figure 9.8   From intended to realised strategy.

As mentioned at the start of this chapter, the improvements that the supple systems approach can generate can be enhanced if linked to a strategy for facilities provision. This provides an overall thrust for the individual actions and especially for the prioritising of actions.

This relationship between an overall thrust and individual initiatives is well illustrated in Figure 9.8 (Mintzberg and Waters, 1985). This figure stresses that the intended strategy is an overall direction, which may be discarded in part in the light of events (unrealised), which may be diverted somewhat by individual initiatives (emergent) but *still defines the overall direction*. The next section turns to strategy formulation in this context.

9.4   Strategic facilities management

9.4.1   *Possible relationships between facilities management and strategic planning*

It is critical that facilities management and corporate strategic management mesh. It has been suggested that there are four possible relationships that could exist between facilities management and corporate strategic planning (Becker, 1990):

- *Administrative linkage*, in which facilities management provides day-to-day operating support, but is itself relatively unimportant in the planning process.
- *One-way linkage*, in which facilities management largely reacts to corporate strategic initiatives. This is the most typical relationship and the one that many facility managers would like to consign to the dustbin!
- *Two-way linkage*, in which there is a reciprocal and interdependent relationship between facilities management and the corporate strategic planning

process. Here facilities management is viewed as credible and important. It is proactive and fully involved in helping guide the development of strategic plans. For example, the facilities manager would be asked to evaluate potential acquisitions and help plan their integration into existing facilities.

- *Integrative linkage*, the highest level of integration in which there is a dynamic, ongoing dialogue, both formal and informal, between the facilities management planners and corporate planners. At this level the facilities manager would be involved in all strategic business decisions, even those that do not directly concern the facility function.

The case studies in Chapter 1 would appear to back up the view that the second option is probably the most common at present. The benefits of involving the facilities manager in strategic decision making have also been highlighted on a number of occasions throughout Chapter 1. So why are so many organisations neglecting a major source of expert knowledge that could easily be utilised? A number of possibilities are discussed below.

9.4.2 *Factors preventing inclusion of facilities management in strategic planning*

Management structure of the organisation

As the case studies show, facilities managers are rarely high up within organisational hierarchies. They tend to be located at the second or third management level; hence many facilities managers find it difficult to influence corporate decision making in any way. In only very few companies are facilities managers on the board and thus in a good position to fight for the inclusion of facilities issues in the strategic plan. Even when facility managers are at quite a high level, this does not necessarily indicate that they will have equal power or influence as other staff at the same level. This may be because as a non-core service facilities management is often viewed as expendable.

Organisation's understanding of facilities management/property

Owing to the fact that facilities management as a profession is still relatively new, there is a certain amount of mistrust and misunderstanding of what it is about. Support of senior management is therefore an essential factor that can contribute to the influence that facilities management can have. Thus when facilities issues are properly understood by senior management, it is likely that facilities managers may become more involved in strategic planning. At present, upper-level managers often take a short-term view of property issues; for instance, maintenance budgets may be one of the first to be cut in times of hardship. These executives fail to see that small savings in the short term may lead to greater expenditure later.

Senior executives may also feel that organisations are continually changing and thus it doesn't make sense to plan too far into the future; however, because of the lead times involved, property decisions often do have to be taken quite a while before the situation becomes critical. For example, if an organisation decides to manufacture a new product, as happened in one of the case studies (Case Study 1 in Chapter 1) new premises may be required. Hence, the organisation should consider their options sooner rather than later, otherwise they may end up having to refurbish an existing building when it would have been better to have a new one designed specifically for the new product.

It may be some consolation to facilities management practitioners to note that the BIFM Survey of Facilities Managers Responsibilities (British Institute of Facilities Management, 1999) provided evidence that a clear majority of members believed that their board of directors, or equivalent, considered facilities management to be of medium or high strategic importance to their organisation.

Facilities managers' understanding of organisation's objectives

In a similar vein, facilities managers do not always have a clear understanding of the core business and hence they are left out in the cold when important decisions are made. It is therefore essential that facilities managers take the time to learn what the core business is really about. Without this understanding, it is impossible for facilities departments to be more proactive. If facilities managers are unable to take the initiative, senior managers may conclude that they are happy to remain in a reactive mode. Thus facilities managers should recognise the need to provide high-quality, proactive and cost-effective services to maintain credibility with their client base.

Staffing/structure of the facilities management department

As would be expected, many facilities professionals are from a construction background, such as architecture, surveying and electrical/mechanical engineering. Consequently, many have had very little training in general management and often find it easier to focus on the technical aspects of facilities management, rather than the people aspect. Hence, a facilities manager who does not appear to manage the facilities department very well, will probably not be viewed favourably by the core business.

The structure of the facilities department in relation to the rest of the organisation is also a critical factor. Many facilities departments have not really been planned and have therefore developed in a haphazard fashion. Consequently, not all building-related functions are grouped together under facilities management and so strategic building decisions may be the responsibility of a separate department. For example, in one of the case studies (Case Study 3 in Chapter 1), the property division is responsible for strategic issues, whilst the facilities department is purely operational. In contrast, some organisations have found that it makes sense to merge their different property sections, so that all building-related decisions and information is grouped together, as in the case of the hospital group (Case Study 5).

9.4.3 *Facilities strategy*

In simple terms, the aim of strategic facilities management should be to achieve a strategic fit between core business needs and the provision of facilities management. However, the above factors, backed up by the case studies, illustrate that in reality there may be a huge chasm of misunderstanding between the two groups and thus many facilities departments are forced to remain in a reactive operational mode. Obviously such problems cannot be conquered overnight and changes are unlikely to happen unless the facilities team can demonstrate how a facilities strategy, designed to support the core business strategy, could benefit the organisation. It may be that the facilities team will have to formulate a strategy, discussing the implications at various stages with key members of staff within the core business, in order to gain their support. Then hopefully the facilities team will be in a position to approach senior management and demonstrate why the facilities strategy should be considered alongside the core business strategy. It can be seen that this can be a natural progression from the relationships and information sources developed through the supple systems approach already described in Section 9.3 of this chapter.

So how should a facilities department go about producing a facilities strategy? Obviously organisations vary substantially and so will their strategies, but the facilities manager may find it helpful to follow the process illustrated in Figure 9.9 (Argenti, 1980).

Objectives

The first stage of any strategic planning process is to determine the objectives of the organisation. In the case of a facilities strategy this will mean interacting with

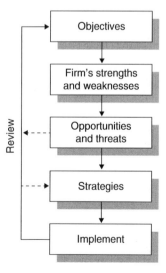

Figure 9.9   Strategic planning process.

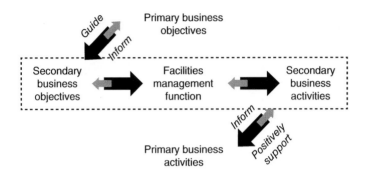

Figure 9.10   Strategically integrated facilities management.

the core business to see what its objectives are and what changes are likely to occur over the next few years. As stated earlier, many facilities departments do not really understand what the core business is about and so this stage may take a while. This nesting of the secondary facilities management objectives within the context of the primary objectives of the core business is crucial. Figure 9.10 illustrates the connections that initially may flow down the diagram, but that will increasingly become two-way communications as the strategic capacity of facilities management increases.

Only when the facilities group understands what the organisation's objectives are will it be able to identify how it as a group will need to change to meet the challenge ahead. In doing this, a five- to ten-year perspective should be taken, to draw consideration beyond a purely financial view.

SWOT analysis

Having identified what the organisation intends to do, the next stage is focused on the facilities department itself and involves analysing the group's *internal* strengths and weaknesses and then the *external* opportunities and threats that it may encounter in the future. 'External' in this context would include the core business. This element of the process is becoming increasingly well known and is commonly referred to as a SWOT (**S**trengths, **W**eaknesses, **O**pportunities, **T**hreats) analysis. The aim is to produce a relatively short list of the major factors that the department ought to take into account when arriving at its strategies. For internal facilities departments market testing is often seen as a threat. For consultants it is a great opportunity.

The reader may find it useful to consider the points in Table 9.2 when compiling a list of major factors. Typically a long list of all possible factors is created first, often involving a wide range of people. This is distilled down to the, say, five major influences in each of the four areas of strengths, weaknesses, opportunities and threats. This small group of factors is the basis from which the creative phase of strategy formulation can be started.

Table 9.2   SWOT analysis.

Internal strengths/weaknesses	External opportunities/threats
People	Political (legal)
Economics (finances)	Economics
Structure	Social
Technology	Technical

Strategies

The facilities group should now know what it is trying to achieve, what sort of shape the department is currently in and what external forces pose threats or provide key opportunities to the department in the longer term. The next stage is to identify what changes will need to be made within the department, taking all of the above factors into consideration, so that the organisation's objectives can be realised. A good way of doing this is for the facilities group to imagine an ideal world scenario, which sets out what the facilities department should be like in the future and then compare this with the current situation.

When the facilities group is assessing what changes will need to be made, there are a number of areas that they should take into consideration:

- *Buildings/space.* Is the existing building stock compatible with the proposed changes? If not, would it be better to construct a new building, refurbish an existing one or move elsewhere? The facilities group may find it useful at this stage to conduct a property audit, such as that conducted by the school (Case Study 2), to ascertain what the current property situation is and compare this with the required future property to determine where action will be necessary. This comparison should consider not only the amount of property, but also its location, size, appropriateness for use, state of repair, maintenance and running costs, tenure and lease terms, and general quality. From this analysis, a strategic plan for the development of the property can then be established.
- *Deployment/structure.* This relates to the choice of the functional units. Are new services likely to be required within the department in the future in order to meet changing needs? For example, it is likely that the organisation will need to carry out a lot of building work? If so, the organisation may wish to expand its facilities services to include an in-house project management team, so that the organisation has more control and is better informed about new building work. In a similar vein, would the organisation benefit if certain existing services were merged with the facilities department to provide a more integrated service, such as occurred in the case of the hospital group (Case Study 5).
- *Resources/staff, technology, finance.* What changes will be required in the way of resources? Are existing facilities staff capable of meeting the challenges

ahead? If not, should they be retrained or would it be better to take on new employees? Will the organisation's/facilities department's existing information technology be able to cope with the changes or will new computers, etc., be necessary? Finally, the financial implications of all the above changes will have to be considered.

- *Contracting out.* In principle, what balance does the organisation want to strike between in-house and contracted facilities management services? Chapter 4 examines this issue in detail.

Once the strategies for each of these individual areas have been worked out, it is important to check that they are mutually supportive, and the facilities team will have to combine the various points to ensure that the greatest collective impact is achieved.

Scope of strategic planning

In recent years the scope of strategic planning has broadened and deepened to include a range of other issues, notably those of the environmental impact of facilities and ensuring business continuity. Given their importance, an outline of these two areas is provided in the following paragraphs.

Just as facilities affect the environment, so external forces can disrupt business continuity. Risk and uncertainty are part of the everyday operating environment of an organisation. Occasionally the risk may result in an impact that may be of sufficient scale to generate a crisis which, if left unattended, may result in severe damage or disaster. A significant interruption to the continuity of operation of an organisation may mean reduced revenues, lost customers, damaged reputation or a significant breakdown in public-service provision. All this would be in addition to material or physical loss or damage that may have occurred. A major interruption can threaten an organisation's survival or its ability to meet statutory or social obligations.

The risk to an organisation may not necessarily be a consequence of a notable disaster incident but may be a result of a badly managed business process, which can result in adverse stakeholder response or a threat to public safety or comfort. Planned disturbance to corporate activity, such as a comprehensive relocation exercise or a major building refurbishment, whilst retaining occupation, could result in severe operational disruption if ineffectively managed.

Business-continuity planning is a managerial obligation that should not be overlooked or postponed because the likelihood of experiencing a major adverse event is remote. A business-continuity process should be designed to enable organisational recovery to proceed swiftly and smoothly to limit the damage caused and to restore operational normality. Business-continuity planning may be identified as a four-stage process:

- *Prevention*, actively reducing the likelihood of a risk occurrence by a high level of risk awareness and prudent planning.

- *Preparedness*, developing an organisational culture of readiness to respond in a controlled and effective manner.
- *Reaction*, implementing a disciplined and regulated form of response that is appropriate and timely in respect of the nature and scale of the incident.
- *Recovery*, activating planned procedures to minimise the degree of damage or disturbance and returning the organisation as close as possible to its pre-incident status.

Not all risks are necessarily insurable, particularly less tangible aspects such as loss of reputation or customer confidence; therefore the need for loss control and risk management is highly significant and quite likely to be a part of the duties of the facilities management function.

Major disaster events may be speculated upon although their likely occurrence and form may be remote and difficult to identify. Such events produce unique sets of circumstances that may be outside the experience of the operational management, resulting in an ad hoc response with varying degrees of successful outcomes. Nevertheless, in a well-managed organisation a structured strategic approach will produce substantial benefits in the form of a level of preparedness that would not otherwise have been present, together with a set of response procedures that are capable of adjustment to meet the specific demands of a particular incident.

The rapid pace of change within commercial and public service sectors, and the greater uncertainty that new ways of operating bring, has raised the degree of risk exposure and created the requirement that effective risk management should come within the scope of organisational management. A considerable portion of risk-bearing activities are often found in the domain of the facilities management function for which practitioners need to be adequately prepared.

Performance evaluation

Facilities management is based on the premise that the effectiveness and efficiency of an organisation is linked to the physical environment in which it operates and that the environment may be improved with direct benefits in operational performance. The objective of facilities management is not limited to optimising running costs of buildings, although that is important, but encompasses the design and management of space and related assets for people and processes in such a way as to support the achievement of the organisational mission and goals in the best combination of suitability, efficiency and cost.

Performance, in operational terms, means the manner or quality of functioning and relates to a building's ability to contribute to the fulfilment of the functions of its intended use. Facilities represent a substantial portion of the assets of most organisations and their operational costs; therefore it is no surprise that an awareness of the performance of these assets in the form of a structured appraisal mechanism is receiving considerable attention.

Performance measurement allows an organisation both to derive information from the past and to evaluate contemporary trends in a manner that will provide the means for better planning and operation in the future. Defining performance measures allows an organisation to:

- establish its position through the careful and consistent evaluation of performance;
- communicate its purpose and direction through targeting what it seeks to achieve, by what means, and to which timescale;
- stimulate action through identifying what is to be done, who is required to act and in what manner;
- facilitate learning throughout the organisation by explaining the levels of achievement attained;
- influence and modify behaviour to promote consistency with operational goals.

Implementation

By this stage the facilities team should have a workable facilities strategy that has been developed with the cooperation of key staff from the core business. Consequently, the facilities team should now be in a position to approach senior management with their ideas. It is probably highly unlikely that senior management will agree in the first instance to every proposal and so facilities managers should not be disheartened if they are not welcomed with open arms. As stated earlier, it may take years before facilities strategies are viewed as an integral part of the organisation's strategic planning process. However, if the facilities team have a fully researched and coordinated strategy, senior management will hopefully be persuaded to allow some of the suggestions to be put into practice.

Consequently, the facilities team will now be at the implementation stage. Thus it will be necessary to identify individuals within the department who will be responsible for pursuing matters on a day-to-day level. A mechanism that can assist in this respect is to request that those identified for a given area provide a brief review of that area and also provide specific targets for the next year, plus an outline of future actions. In this way the strategies become owned at operational level and thus the department should make definite progress towards achieving the required changes.

Review

The setting of specific individual targets means that a control mechanism is now in place that allows the targets to be periodically and systematically reviewed. Thus the next stage is to establish a monitoring system to track whether the targets are being achieved and whether this is being done in the most effective and efficient manner. It is important to remember, however, that the targets are being pursued for the benefit of the whole organisation and so the facilities group must ensure that their activities remain relevant to what is occurring in the core business. If necessary, the department may have to reassess the situation and revise its targets, or even rework its strategy entirely.

Summary

The above discussion should give the facilities manager some idea of how to address the subject of strategic facilities management. The reader should try to remember that senior management may not be very receptive at first to the idea of developing a facilities strategy alongside the core business strategy, but it is worth persevering as the potential benefits to the organisation could well be substantial, as the case studies demonstrated.

The other key benefit is the integrating focus it can create for the initiatives generated within the facilities management department. The outcome of these initiatives will of course depend on the *individuals* involved and the next section considers this specific aspect in more detail.

9.5   Learning organisations

9.5.1   *Context*

The ultimate objective of any organisation (or department) must surely be to achieve self-managing workers who are motivated to achieve high quality, be capable of achieving high quality and able to exhibit self-control. To the extent that this can be achieved there will be little need for systems, nor for leaders in the traditional sense. A current school of thought has suggested that such objectives can be attained by creating 'learning organisations'. This section will discuss how learning organisation principles could be utilised by facilities managers.

In order to become a learning organisation the majority of organisations would require a radical change in the way people behave and, consequently, the transition is bound to take several years. However, once an organisation has encompassed truly sustainable organisational learning principles, they will undoubtedly have achieved a substantial competitive advantage. Bearing this in mind, it is obvious that organisational learning is a complex subject and as such cannot possibly be summarised successfully in a couple of pages. The following discussion is meant only as an introduction to the subject and readers should consult the references and Appendix to this chapter for further information.

9.5.2   *Individual learning*

Organisations learn only through individuals who learn. Individual learning does not guarantee organisation learning. But without it no organisational learning occurs (Senge, 1990).

Thus the first step on the way to becoming a learning organisation is to encourage individual learning. The key to individual learning is to expand our awareness and understanding by questioning and challenging our actions and assumptions.

There are two goals in any learning process. One is to learn the specifics of a particular subject and the other is to learn about one's strengths and weaknesses as a learner. However, people are very different and learning styles are similarly unique. Thus people should try to work out the learning method that best suits them. Just to give an indication of the possibilities, they could adopt one of the following methods: action learning, on-site and off-site courses, self-development programmes, distance-learning systems, coaching and counselling, job rotation, secondments and exchange programmes.

Even though people may express the desire to learn, they may often be hampered by their 'mental models' (Senge, 1990). Mental models are deeply ingrained assumptions/generalisations that influence how we understand the world and how we take action. Consequently, when people make decisions they not only consider all the facts but also unconsciously refer to their mental models. Individual learning requires that we try to uncover our mental models, so that we can see how they affect our actions and prevent us from learning.

The *left-hand column exercise* is a technique that allows people to 'see' how their mental models operate in particular situations. The left-hand column exercise helps managers to identify how their mental models play an active, but often unhelpful, part in their management practices. It shows how managers will often manipulate a situation to avoid dealing with difficulties head-on and thus problems will not be solved. After carrying out the exercise, facilities managers should begin to see why they should deal with their assumptions more forthrightly in the future.

To carry out the left-hand column exercise, the manager first selects a specific situation where he is interacting with another person or people in a way that he feels is not working. Then on the right-hand side of a sheet of paper, the manager writes exactly what is said, while on the left-hand side the manager writes what he is actually thinking, but not saying. The following example outlines a typical problem conversation (Senge, 1990):

What you're thinking	What is said
Everyone says the presentation went very badly.	**You**: How did the presentation go?
Does he really not know how bad it was? Or is he just not willing to face up to it?	**Bill**: Well, I don't know. It's really too early to tell. Besides we're breaking new *ground here*.
	You: Well, what do you think we should do? I believe that the issues you were raising are important.
He really is afraid to see the truth. If only he had more confidence, he could probably learn from a situation like this. I can't believe that he doesn't realise how disastrous that presentation was to our moving ahead.	**Bill**: I'm not so sure. Let's just wait and see what happens.
I've got to find a way to light a fire under him!	**You**: You may be right, but I think we may need to do more than just wait.

Table 9.3   Wanted and unwanted behaviours in a learning organisation.

Wanted behaviour	Unwanted behaviour
Asking questions	Acquiescing
Suggesting ideas	Rubbishing ideas
Exploring alternatives	Going for expedients, quick fixes
Taking risks/experimenting	Being cautious
Being open about the way it is	Telling people what they want to hear/ filtering bad news
Converting mistakes into learning Reflecting and reviewing Talking about learning	Repeating the same mistakes Rushing around keeping active Talking anecdotes (i.e. what happened, not what was learned)
Taking responsibility for own learning and development	Waiting for other people to do it
Admitting inadequacies and mistakes	Justifying actions/blaming other people or events

The most important lesson to be drawn from the above example is how people undermine their opportunities for learning when faced with a difficult situation. Rather than dealing with the problem head-on, the two people talk around the subject. Consequently, the problem is not resolved and there is no clear way forward. There is no one 'right' way of dealing with difficult situations such as this. However, the left-hand column enables managers to see that their actions may actually be making a situation even worse. The next stage is to try to work out how the situation could be improved, so that managers and their staff can both learn from the experience.

One method of doing this is known as 'balancing inquiry and advocacy' (Argyris, 1982). The use of pure advocacy encourages people to argue, putting their opinion ever more strongly just to win the argument, without really considering other people's viewpoints. Similarly, pure inquiry is also limited; just asking lots of questions can be a way of avoiding learning, as people may do so to avoid having to put their own opinion forward. It therefore makes sense to combine advocacy and inquiry, so that everyone makes their thinking explicit and subject to public examination. In this way people can begin to question their own mental models through discussion with others and discover that there are alternative, perhaps better, options.

The above discussion highlights the fact that individual learning involves taking risks. If managers want to learn from situations, they must be prepared to put their own views on the line and be prepared to admit that they were wrong. In a similar vein, managers must allow staff to make their own mistakes (within safe limits), so that they can also learn from them (Byham, 1988). However, many individuals may feel that mistakes will result in punishment of some form and so the facilities manager should take steps to ensure that staff are fully aware that this is not the case. In essence, facilities managers should recognise the need to foster environments that encourage individual learning, making clear the sort of behaviour that they would like to see, such as that listed in Table 9.3 (Honey, 1991).

9.5.3   Team learning

Individuals learn all the time and yet there is no organisational learning. But if teams learn, they become a microcosm for learning throughout the organisation.

Skills developed can propagate to other individuals and to other teams. The team's accomplishments can set the tone and establish a standard for learning together throughout the organization (Senge, 1990).

As far as team learning is concerned, the goals remain the same as those of individual learning, namely learning about the specifics of a subject and learning about one's strengths and weaknesses as a team learner. With regard to the learning process, the learning cycle may actually work better in the case of a team than with an individual. This is because individuals cannot realistically be expected to be equally skilled in each stage of the process and as a consequence may well take shortcuts, whereas a team containing individuals who are good at different stages of the process will have the potential to learn well. However, unless the individuals are actually able to work well together, their individual skills will be wasted.

Learning to work together as a team can be quite difficult. This is because the needs of the individual, as well as those of the team, will come into play. Belbin's (1981) research into team learning has discovered that successful teams are composed of members offering a wide range of team roles and a spread of mental abilities. Surprisingly, teams composed wholly of individuals with high mental abilities did not perform very well, as the members tried to compete with each other. Table 9.4 lists the major team roles that are required for an effective team.

It should be noted that different roles need not each represent an individual; one person can perform more than one role. In reality, of course, teams are not always composed of people with these complementary skills. Consequently, if teams find they do not have the ideal distribution of talents, they may find it useful to identify where gaps exist and assign specific team members to cover the missing roles.

As well as being better able to accomplish tasks jointly there are, of course, many opportunities for the individual to learn from working in a team. Creating different groupings for different assignments is thus a way by which individual learning can be spread more widely within the organisation. To show how some of these factors combine in a practical situation, a brief case study is presented next.

Table 9.4   Belbin's team roles.

(1) *Chairman* ensures that the best use is made of each member's potential. Is self-disciplined, dominant but not domineering.

(2) *Shapers* look for patterns and try to shape the team's efforts in this direction. They are out-going, impulsive and impatient. They made the team feel uncomfortable, but they make things happen.

(3) *Innovators* are the sources of original ideas. They are imaginative and uninhibited. They are bad at accepting criticism and may need careful handling to provide that vital spark.

(4) *Evaluators* are more measured and dispassionate. They like time to analyse and mull things over.

(5) *Organisers* turn strategies into manageable tasks which people can get on with. They are disciplined, methodical and sometimes inflexible.

(6) *Entrepreneurs* go outside the group and bring back information and ideas. They make friends easily and have a mass of contacts. They prevent the team from stagnating.

(7) *Team workers* promote unity and harmony within a group. They are more aware of people's needs than other members. They are the most active internal communicators and cement of the team.

(8) *Finishers* are compulsive 'meeters' of deadlines. They worry about what can go wrong and maintain a permanent sense of urgency which they communicate to others.

(9) *Specialists* do not have a recognised team role, but will often be a necessary part of a team as they possess unsurpassed technical knowledge and experience.

Case Study: An organisation learning to learn

The organisation is well known for being at the forefront of facilities management developments. The problems it encounters are a reflection of its willingness, indeed desire, to achieve constant improvements. Their theory-in-use is 'you do not have to be sick to get better'.

The provision of facilities management services was carefully studied by various people within the facilities group and it was concluded that cultural change was required to achieve the next level of service quality desired. Various barriers to this change were identified, which resulted from the organisation context and past experiences of the people involved. These are summarised in Table 9.5.

Despite these barriers, the facilities department decided that change was necessary. Hence during 1991–1992 a TQM programme was implemented based on previous experience and focusing on: management commitment, focus on teams, continuous improvement activity, training, empowerment, use of quality tools, recognition of performance and customer orientation. A successful workshop was held at the start of the process and a lot of enthusiasm was generated. A range of process-improvement teams was created, which identified and documented 91 processes and 487 process-improvement actions.

Unfortunately, the volume of work involved resulted in widespread demotivation and so the activities were synthesised around seven key

Table 9.5 **Barriers to cultural change.**

Factor	Description
Cultural	The 'fixer' is encouraged by the organisation. Facilities management organisations have grown from the caretaker and servant of the firm. Recognition by senior management in the facilities management organisation, as well as the core business, was given for reactive solutions to problems. Retribution would be given if you questioned the way things were done. Subordinates were out of line if they suggested that there was a better way of doing something to their senior management.
Hierarchy	Structures of organisations provide levels of power within departments and across departments, e.g. juniors must not speak to seniors in other departments. This is compounded by the view that internal service departments have been expected to provide 'services' to the management of the core business. Individuals at all levels in the facilities management and core organisation have built their power base on this formality of communication.
Processes	Processes have evolved over time, built up from convenience, compromise, rituals, past experiences, existing departmental structures and established responsibilities. They are tried and tested, they are seen to work, and thus are best left alone.

processes. Some good progress was made, but the overall effect was patchy. It was concluded that the use of the TQM tools described was *not* producing sustainable change based on improved teamwork and responsiveness to customers.

Consequently, from March 1993 the emphasis shifted to interventions designed to produce a learning organisation. This was and is to be based on effective 'learning individuals'. Thus the organisation is endeavouring to build on its TQM experience, but is emphasising a cultural change that revolves around individuals at all levels becoming adept at more systemic thinking than is usual. In this way individuals will be addressing issues, inventing solutions with a much richer palette – looking well beyond the confines of their local situation. The objective is to push back people's 'learning horizons'. At a low, but important, level, an example was the tea boy, *himself*, suggesting that he took the mail round.

Another key development is shifting the emphasis from facilities managers as fixers to facilities managers and their staff as thinkers and planners. For example, they will now not accept a brief from a client that simply says 'I want to be able to retrieve any file within one hour.' Instead they ask, 'Why do you need that level of service? Do you realise how expensive it will be?' etc. This has caused some initial friction, but as benefits flow from a more careful analysis based on the client's real needs, the facilities staff are becoming more confident in this role of analyst, and the clients are beginning to value it as a valuable part of the service on offer.

As stated earlier, individual learning is necessary, but not sufficient to create a learning organisation. The next building block is the operational team. Consequently, in the facilities department, teams are now being assessed against Kolb's learning cycle (Kolb, 1976) (Figure 9.11), with the objective of making sure the full range of problem-solving orientations are present. They are also studying how the connections between the quadrants can be made to operate effectively, so that the learning cycle can genuinely spin from internal sensing of an issue, to careful observation (watching), to thinking and then to action (doing). This is being linked to consideration of Belbin's team roles.

It is known that to be creative an organisation must be tolerant of errors. This problem has been recognised with the facilities department and so, within safe limits, the facilities manager consciously lets teams make what he perceives are mistakes. Very often he is right, but the growth in learning capability of the team justifies it. Sometimes the team is triumphantly right and their motivation soars.

Comment

One of the key lessons from this case is that an extra dimension is added to the facilities manager's decision making. He may decide to do things in a

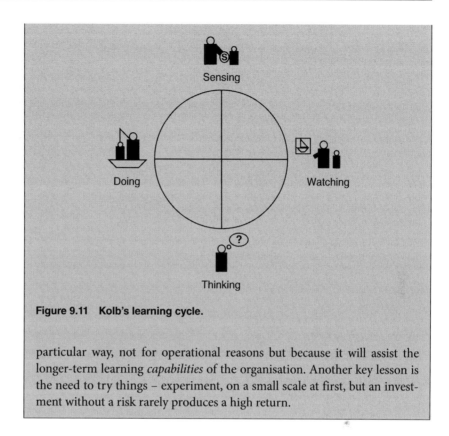

Figure 9.11   Kolb's learning cycle.

particular way, not for operational reasons but because it will assist the longer-term learning *capabilities* of the organisation. Another key lesson is the need to try things – experiment, on a small scale at first, but an investment without a risk rarely produces a high return.

9.6   Conclusions and interactions

9.6.1   *The changing FM landscape*

Our continued attempts to routinise and standardise facilities management are fraught with difficulties. Undoubtedly, innovations and coordinated international efforts have secured a common language for the profession – a language that has enabled the ready exchange of FM services. This in turn has given rise to two key developments: (1) the ability to measure and evaluate competing services and (2) the ability to interoperate and collaborate with numerous partners in a much more fluid and responsive manner. However, standardisation can also create a 'straitjacket' that stifles our response to change both within the internal environment organisations and the external environment. It is this ability to respond to such changes that characterises excellence in facilities management. This may involve tearing up the rulebook and starting again.

The 'systems' approach described in Chapter 1 (and elaborated on in Chapter 7) provides a consistent approach that constantly reminds us of the unpredictable nature of the FM environment. It moves us from a position of 'comfort' to one of ongoing 'vigilance'. It incorporates key realities in the FM environment that are all too often 'smoothed out' by IT solutions that seek to eliminate decision making on the ground.

Look around at the turmoil that continues to affect financial markets, climate change and major demographic changes (e.g. the advent of the greying society). Added to this are other unforeseen changes that will continue to impact on the way in which facilities are procured and managed. Resilience becomes the watchword. The management approach becomes not one of 'getting it right first time' but being fleet of foot to allow responses to unpredictable developments. This involves having a flexible asset base (property portfolio) and an interconnected understanding of technology, culture and organisational activities.

In the spirit of 'responsive' decision making, which underlies 'systems theory', this book serves to reinforce our approach to adaptive change. Whilst embracing the value of standardisation it also recognises that it should not be blind to external change. Recognition of this fact is acknowledged even by one of the greatest proponents of standards, Henry Ford, who said:

> If you think of standardization as the best that you know today, but which is to be improved tomorrow; you get somewhere (Henry Ford, American Industrialist, 1863–1947).

9.6.2 *Where we started*

In Chapter 1 the intention was to encourage facilities managers to take a second look at how they manage their facilities. Because the landscape of facilities management is so diverse, the idea of providing a 'how to' recipe for universal application is misleading at best. We saw that facilities management operations can be categorised in terms of location. This can range from an FM operation involving only a single office manager, a single site operation, a localised site operation, an FM operation involving multiple sites – right up to an international FM operation. Each scale presents its own challenges. For the single office manager, the problem of organisational change demands an agile response. Overseeing the fit-out of a new office floor or accommodating new ideas in new ways of working can be fraught with peril. Applying an inappropriate FM solution can quickly undermine the performance of a small start-up company. Dysfunctionality caused by facility design can be all too apparent. Contrariwise, successes in FM at the local level are highly visible and many opportunities arise for experimenting with radically different approaches (tempered with considerations of organisational risk).

At the other end of the spectrum, multiple sites and international operations present different challenges. Performance measurement becomes a critical task if a facilities manager is to successfully compare operations on a global level. Inevitably, the consequence of this is that commoditisation and common standards become indispensable. Set against this is the need to encompass cultural diversity and local practices that more accurately reflect the regional or national context. The tension between standardisation and local practice extends to IT implementation, knowledge management, workplace design and building technology.

One of the unique features of facilities management is 'asset specificity'. This rather grandiose term describes a very simple situation: every building is different. Each building differs in terms of the history of the physical asset, the type of organisation it accommodates and the experience and expertise of the local FM team. Avoiding comparisons of 'apples with oranges' is imperative. Consider, for example, the comparison of energy performance within a property portfolio. Buildings with an age in excess of 100 years are likely to demonstrate energy performance characteristics very different from buildings of only 10 or 20 years in age. Not only is the absolute energy performance likely to differ significantly but more importantly the ability of such facilities to incorporate energy retrofits will inevitably differ markedly.

Many FM subunits in an organisation (e.g. floor operations, site management, regional management) can give rise to significant levels of 'recursion' such that the viability of the FM system is intractably linked to the viability of independent subunits. A high level of 'autonomy' is often exercised by different FM subunits within an organisation, allowing a localised and varied response to changes in the external environment. It recognises how variety in the external environment needs to be matched by variety in the internal FM operation in order to respond to change.

In Chapter 2 the way in which space impacts on our sensations and perceptions was explored. It highlighted the abundance of new research that is forcing us to reconsider our assumptions about space and its capacity to influence our behaviour. Light affects the way that we perceive spatial size, temperature, the size of objects and sounds. Other factors such as floor layout, acoustics and air quality affect our behaviour in unexpected ways. Armed with knowledge of how these factors impact on individuals and groups, facilities managers are able to explore many different ways of achieving the same ends. Instead of relying on simple physical measures to establish phenomena such as spaciousness, we are able to consider many other factors that affect our experience. Unlocking our understanding of perception opens doors to many facilities management interventions that are not tied to costly and resource-intensive activities. Instead, the focus becomes that of the human mind rather than simply the physical asset. Nowhere is this more apparent than in school design. Subtle and often unintuitive design decisions can significantly affect learning outcomes in schools. Indeed, the chapter demonstrates how we can set about measuring the impacts of specific design decisions on learning outcomes. For many in FM, this has been a holy grail in performance measurement, whereby qualitative factors that centrally affect the performance of the client organisation become measurable phenomenon. No longer is the facilities manager banished to considerations on the left-hand side of the balance sheet (cost reduction) but can also more importantly assess the impact of facilities in relation to the right-hand side of the balance sheet (increases in effectiveness).

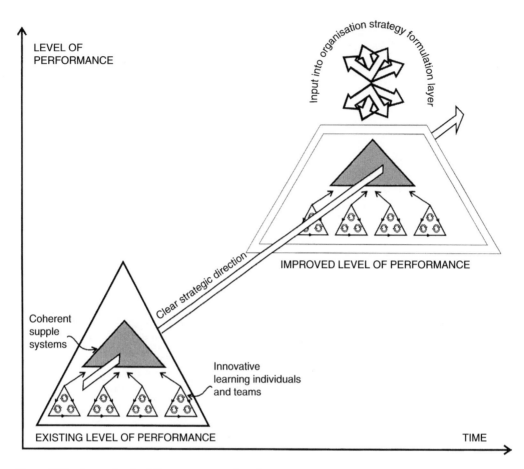

Figure 9.12   Improving facilities management performance.

9.6.3 *Finally*

Picking up on the promise and challenge of progressively striving for excellence in facilities management through a strong and effective user orientation, the middle chapters of this book have addressed various specific areas. These have covered issues of: organizing for FM, dynamically linking to and communicating with stakeholders and utilizing various technical and process tools with finesse.

Against this rich variety of possibilities, this last chapter has focused on organizational ways of progressively *improving* facilities management performance over the medium term. The thrust of the argument we are putting forward is that an organizational environment must be created in which people know what needs to be done to support the core business better. This is achieved through feedback *information* within supple systems, *direction* provided by a strategy linked to corporate aspirations and, underpinning these mechanisms, actions to create learning individuals and groups, within an organisational context that encourages them to *innovate*.

It should now be apparent that the full effect is dependent on all three aspects working in unison, but it has been argued that in general the feedback information is the first place to start as this will inform strategy formulation and begin to create a better-informed customer orientation amongst staff. If, for a given organisation, say, a strategy already exists then this can be the springboard for the other areas instead. The particular order is not critical.

The view presented here can be summarised as shown in Figure 9.12. The improved level of performance is the product of innovations, at all levels of the facilities management organisation, each orientated towards clearly stated strategic goals by integrative supple systems. The end result of this enhanced dynamism is the creation of a continuously learning and innovating facilities management function that meaningfully contributes to organisational strategy formulation.

To address all three areas successfully will take an organisation several years and involve all aspects covered earlier in this book, but the prize is dynamic, continuous improvement in line with the needs of the core business. With this comes satisfied clients and, therefore, sustainable success and security for the facilities team.

References

Attwood, M. and Beer, N. (1990) Towards a working definition of a learning organisation. In *Self-Development in Organisations*, (eds M. Pedler, J. Burgoyne, T. Boydell and G. Welshman). McGraw-Hill, New York.

Argenti, J. (1980) *Practical Corporate Planning*. Allen and Unwin, London.

Argyris, C. (1982) *Reasoning, Learning and Action: Individual and Organizational*. Jossey-Bass, San Francisco, CA.

Barrett, P. (1989) Quality assurance in the professional firm. In *Quality for Building Users Throughout the World*, **3** volumes, Vol. I. CIB. pp. 181–190.

Barrett, P. (1994) *Supply Systems for Quality Management*. RICS Research Paper Series, RICS, London.

Becker, F. (1990) *The Total Workplace – Facilities Management and the Elastic Organization*. Van Nostrand Reinhold, New York. p. 81.

Belbin, R. (1981) *Management Teams: Why They Succeed or Fail*. Heinemann, London.

British Institute of Facilities Management (1999) *Survey of Facilities Managers' Responsibilities*. BIFM, Saffron Walden.

Byham, W. (1988) *Zapp – The Lightning of Empowerment*. Century Business, London.

Easterby-Smith, M. (1990) Creating a learning organisation. *Personnel Review*, **19**(5), 24–28.

Gronroos, C. (1984) *Strategic Management and Marketing in the Service Sector*. Chartwell-Bratt, Bromley.

Honey, P. (1991) The learning organisation simplified. *Training and Development*, July, 30–33.

ISO (1991) *ISO 9004-2: Quality Management and Quality System Elements – Part 2: Guidelines for Services*. International Organization for Standardization, via British Standards Institution, London, as Part 8 of BS 5750.

Jashapara, A. (1993) Competitive learning organisations: a way forward in the European construction industry. In *Proceedings of CIB W-65 Symposium 93: Organisation and Management of Construction – The Way Forward*, Trinidad, September 1993.

Kast, F. and Rosenzweig, J. (1985) *Organization and Management: A Systems and Contingency Approach*. McGraw-Hill, New York.

Kolb, D. (1976) *The Learning Style Inventory Technical Manual*. MacBer, Boston, MA.

Louis, M. and Sutton, R. (1991) Switching cognitive gears: from habits of mind to active thinking. *Human Relations*, **44**(1), 55–76.

Mintzberg, H. and Waters, J. (1985) Of strategies deliberate and emergent. *Strategic Management Journal*, **6**, 257–272.

Mumford, A. (1991) Learning in action. *Personnel Management*, July, 34–37.

Nonaka, I. (1991) The knowledge creating company. *Harvard Business Review*, November/December.

Pedler, M., Burgoyne, J. and Boydell, T. (1991) *The Learning Company*. McGraw-Hill, New York.

Revans, R. (1982) The enterprise as a learning system. In *The Origins and Growth of Action*. Chartwell-Bratt, Bromley.

Senge, P. (1990) *The Fifth Discipline: The Art and Practice of the Learning Organization*. Doubleday, USA.

Appendix: Models of learning organisations

The following table is adapted from Jashapara's literature review on learning organisations (Jashapara, 1993) and summarises various approaches to organizational learning.

(1) Learning organisation as five disciplines	
Senge (1990)	*Personal mastery* is seen as developing our capacity to clarify what is important to us in terms of our personal vision and purpose.
	Team learning is seen as developing our capacity for conversation and balancing dialogue and discussion. There can be a tendency in many decision-making processes towards discussion where different views are presented and defended. Senge promotes the greater use of dialogue where different views can be presented as a means towards discovering a new view.
	Systems thinking is seen as developing our capacity for putting the pieces together and seeing wholes rather than disparate parts.
	Mental models is seen as our capacity to reflect on our internal pictures. This discipline involves balancing our skills of inquiry and advocacy as well as understanding how our mental models influence our actions.
	Shared vision is seen as building a sense of commitment in a group based on what they would really like to create. Senge believes that leaders will play a critical role in developing learning organisations especially through building a shared vision which is rooted in personal visions.

(2) Learning organisation as action learning	
Revans (1982)	Revans explores the notion of the enterprise as a learning system. He suggests that such organisations encourage people to regularly study and reorganise their systems of work through learning and the quality of learning is determined by the morale of the organisation. He uses the following equation:
	Organisational learning $L = P + Q$, where P = programmed learning (highly specialist) and Q = questioned learning (asking discriminating questions)
	Such organisations provide 'action learning groups' where managers can learn to take effective action by closely studying and questioning their everyday work. This is in contrast to traditional methods where managers would perform an analysis and make recommendations, but not necessarily take action.
Mumford (1991)	Mumford proposes four 'Is' to achieve effective forms of action learning; *interaction* with major organizational players, *integration* of appropriate skills and knowledge, *implementation* for which managers are personally accountable and *iteration* which views learning as a process. There can be high opportunity costs to action learning and the process is time consuming. An environment needs to develop to create special assignments for action learning which may include exchanges, sabbaticals, counselling and coaching skills.
(3) Learning organisation as encouraging wanted behaviours	
Honey (1991)	Honey's method for creating learning organizations revolves around establishing a number of learning behaviours which need to be encouraged in an organisation and working out suitable triggers and reinforcers for the wanted behaviours. The establishment of the relevant learning behaviours is likely to be critical for sustaining competitive advantage. (Table 2.3 shows possible wanted and unwanted behaviours.)
(4) Learning organisation as continuous transformation	
Pedler, Burgoyne and Boydell (1991)	*Strategy* includes a learning approach to strategy with small scale developments and feedback loops to enable continuous improvement and participative policy making.
	Looking in includes using IT to help individuals understand what is going on and using formative accounting and control to assist learning and delighting internal customers. In addition, this area includes developing an environment of collaboration between internal departments and exploring basic assumptions and values of the reward system.
	Structures implies the need for roles and careers to be flexibly orientated to allow for experimentation, growth and adaptation.
	Looking out includes regularly scanning and reviewing external environment and developing joint learning with competitors and other stakeholders for 'win:win' learning.
	Learning opportunities includes a climate of continuous improvement where mistakes are allowed and encouraged together with self development opportunities for all.

(5) Learning organisation through experimentation	
Easterby-Smith (1990)	Easterby-Smith suggests organisations promote experimentation in various ways: in people to generate creativity and innovation, in structures to introduce flexibility, in reward systems so as not to disadvantage individuals who take risks and in information systems to focus more on unusual variations and to engender an attitude of looking forward rather than living in the past.
(6) Learning organisation as a knowledge creating company	
Nonaka (1991)	Nonaka believes that the lasting source of competitive advantage is knowledge where successful companies continuously create new knowledge, disseminate it widely and embody it in their products. He makes a distinction between tacit and explicit knowledge. Explicit knowledge is systematic and formal and can be easily communicated, whereas tacit knowledge is highly personal 'know-how' and consists of our mental models. It is tacit knowledge which is more illusive and contains our subjective insights, intuitions and hunches.
	Nonaka sees the continual challenge of knowledge creating companies as re-examining what they take for granted. In such organisations, employees are likely to share overlapping information to encourage frequent dialogue and communication. In addition, employees are more likely to be engaged in strategic rotation so that they can understand the business from a wide range of perspectives and have free access to company information. The principle of 'internal competition' is promoted, where numerous groups develop different approaches to the same project and argue over their merits and shortcomings.
(7) Learning organisation models developed through practice	
Attwood and Beer (1990)	In their work on learning organisations, Attwood and Beer suggest a role for management to provide continuous and relevant learning at all levels with the aim of improving organisational effectiveness through individual development. They propose a role for organisational development/personnel staff to offer expertise on a wide range of learning strategies. Such organisations have shared responsibility of staff for the learning process with continuous and regular feedback. For the organisation's survival, the rate of learning is seen as equal to or greater than the change in the environment. These organisations require liberating structures, organisational sub-climates of learning and a culture of responsible freedom.

Index

Facilities Management: The Dynamics of Excellence, Third Edition. Peter Barrett and Edward Finch.
© 2014 John Wiley & Sons, Ltd. Published 2014 by John Wiley & Sons, Ltd.

Printed and bound by CPI Group (UK) Ltd, Croydon, CR0 4YY

18/07/2024

14529827-0001